I0813534

150 GARDENS

YOU NEED TO VISIT BEFORE YOU DIE

Lannoo

Introduction
By Stefanie Waldek

As someone who spends much of my time looking up at the stars, wondering about the mysteries of space, it's funny that I should write a book dedicated to something so intimately bound to Earth: gardens. But if I were to boil my interest in the cosmos down, it's simply that I am enthralled by exploration, both physical and intellectual. That's why I became a journalist—to discover and to share. And that's also why I write about travel, both on Earth and beyond. For me, gardens are the perfect microcosms for exploration here at home, and they're something that can be enjoyed by all kinds of people, all over the world.

I love nothing more than wandering garden paths and being delighted by what I see around every turn. But I'm also amazed by how gardens can transport you anywhere on the planet—and even back in time—through their meticulous designs, magnificent structures, and magical collections. In some gardens, you can trace the paths walked by royalty, philosophers, explorers, and scientists. In others, you can experience gardening as a cultural art form. And in others still, you can simply take in the natural beauty that makes our planet unique. There are countless reasons to spend time in a garden, and each is perfectly valid.

In *150 Gardens You Need to Visit Before You Die*, I've shared a vast range of gardens, from immense botanical institutions with thousands of specimens, to smaller plots for quiet meditations, to museums that combine both artworks and plantings. I hope these brief introductions inspire you to plan a visit or two, whether in your hometown or on your global travels, so that you can enjoy the sights, smells, sounds, and stories of the world's best gardens.

OVERVIEW

OVERVIEW

OVERVIEW

ASIA

OVERVIEW

OVERVIEW

OVERVIEW

OCEANIA

01 ASWAN BOTANICAL GARDEN

El Nabatat Island, Aswan, Egypt

TO VISIT BEFORE YOU DIE BECAUSE

The setting of this botanical garden is highly unusual: an island in the middle of the Nile River.

Standing on the bank of the Nile in Aswan, you'll see Elephantine Island and its many archaeological sites, hotels, and restaurants. But hidden behind it is the less-visited El Nabatat Island, and you'd be remiss not to book a quick ride aboard a felucca to visit it. The island was formerly known as Kitchener's Island for its 19th-century owner, Lord Horatio Kitchener, a British army officer who was gifted it for his services to the Crown. An ardent lover of plants, Kitchener imported flora from around the world to his 17-acre island, laying the foundation for what has since become the Aswan Botanical Gardens. Stroll down the palm-lined paths of this lush oasis, then stop for some tea in the garden's café.

No official website

02 ABURI BOTANICAL GARDENS

Aburi, Ghana

TO VISIT BEFORE YOU DIE BECAUSE

This century-old garden is a respite from the bustle of Accra, the capital of Ghana.

On weekends, Accra's residents make the 45-minute trip into the nearby hills to escape the cacophony and the heat of the city, many ending up at the Aburi Botanical Gardens for a leisurely afternoon. Though by no means the most extensive botanical garden in the world, Aburi offers a range of atmospheres to cater to all sorts of visitors—sometimes it's filled with ebullient picnicking families, sometimes it's enveloped in moody mist. The garden was founded by the British in 1890 as a sanatorium before it transitioned into a botanical garden helmed by Mr. William Crowther of the Royal Botanic Gardens, Kew. Today, highlights include an allée of towering palms, a strangler fig, a "tree of life" with artistic carvings, and, quite unusually, an old helicopter often used by children as a playground.

aburi.ghana-net.net +233 34 219 7473

03 SIR SEEWOOSAGUR RAMGOOLAM BOTANICAL GARDEN

Royal Road, Pamplemousses, Mauritius

TO VISIT BEFORE YOU DIE BECAUSE

The Giant Water Lily Pond looks like it belongs in Jurassic Park.

In 1767, French botanist Pierre Poivre (1719–1786) founded a botanical garden in Mauritius, now called the Sir Seewoosagur Ramgoolam Botanical Garden after the country's first prime minister (though it's sometimes referred to as the Pamplemousses Botanical Garden). Poivre was tasked with acclimating non-native crops to the island, but he brought in non-native flowers, too. The garden's signature feature, its Giant Water Lily Pond, was developed by its second director, Jean-Nicolas Céré (1775–1810), while its 80-species-strong palm collection was initiated by James Duncan, who took up post as director in 1849. Fun fact: Mauritius has only one endemic mammal, the *Pteropus niger* bat, otherwise known as the Mauritian flying fox, which can be spotted in the garden's treetops.

www.ssrbg.govmu.org/SitePages/Index.aspx

+230 243 94 01

04 ANIMA GARDEN

Douar Sbiti, Route d'Ourika, Marrakech, Morocco

TO VISIT BEFORE YOU DIE BECAUSE

This art-filled garden in the shadow of the Atlas Mountains is a sensory delight.

If you tire of the crowds in Marrakech, take a 45-mile drive into the Ourika Valley to visit the eccentric Anima Garden, opened in 2016 by Austrian multimedia artist André Heller (b. 1947). The nearly 5-acre grounds are filled with bright and playful sculptures by international artists interwoven between palm trees, cacti, and bamboo. Walk down one path, and you might find a replica of *The Thinker* by Auguste Rodin (1840–1917). Stroll down another—there's a stele by Keith Haring (1958–1990). Turn a few corners and you might happen upon a tree covered in African masks. It's an eclectic collection, which makes your discoveries as you meander through the garden all the more delightful. And if the works and the flora aren't enough, they're both set against the impressive backdrop of the snow-capped Atlas Mountains.

www.anima-garden.com +212 524 48 20 22

05 LA MAMOUNIA

Avenue Bab Jdid, Marrakech 40040, Morocco

TO VISIT BEFORE YOU DIE BECAUSE

"This is a wonderful place," Sir Winston Churchill said of La Mamounia.

In the 18th century, Alaouite Sultan Mohammed Ben Abdallah (c. 1710–1790) gifted a 32-acre orchard to his son for his wedding. The prince transformed it into a lavish garden for entertaining, and that spirit has been carried through to the current day—the grounds now belong to one of the most famous hotels in Marrakech. Built in 1922 in a combination of Moroccan and Art Deco styles, La Mamounia has welcomed legendary guests, from Winston Churchill (1874–1965), to Charlie Chaplin (1889–1977), to Yves Saint-Laurent (1936–2008) and Pierre Bergé (1930–2017), who would go on to design their own sumptuous garden in the city, Jardin Majorelle. The gardens of La Mamounia are filled with palm trees, cacti, bougainvilleas, delphiniums, and even 700-year-old olive trees, which together create a sensational landscape.

www.mamounia.com/en/the-exclusives/the-gardens.html

+212 (0) 524 388 600

06 SEYCHELLES NATIONAL BOTANICAL GARDENS

Sans Soucis Road, Victoria, Seychelles

TO VISIT BEFORE YOU DIE BECAUSE

Aldabra giant tortoises call this garden home.

While it's true that the Seychelles' main island, Mahé, is already a lush paradise, there are still some plants at the Seychelles National Botanical Gardens in Victoria that you can't see elsewhere on the island, namely a number of non-native species, including some foreign orchids. But the garden is also home to many endemic plants, which are being studied and cultivated by researchers. One of the garden's most famous native plants is the coco de mer, a palm tree that grows the largest seed of any plant in the world (up to 3 feet in diameter, weighing up to 55 pounds). Prince Philip, the Duke of Edinburgh (1921–2021), planted one of the trees in the garden in 1956, and it's still growing strong today. The gardens are also home to a number of Aldabra giant tortoises, some of whom are more than 150 years old.

No official website

+248 4 67 05 37

07 DURBAN BOTANIC GARDENS

9A John Zikhali Road, Berea, Durban, South Africa

TO VISIT BEFORE YOU DIE BECAUSE

These are the oldest surviving botanical gardens in Africa, founded in 1849.

Durban Botanic Gardens has humble roots. Established in 1849, the gardens were originally a living laboratory for the cultivation of produce. Sugar was its most successful experiment—the garden's curator Mark McKen (1823–1872), having worked in Jamaica, knew how to grow the crop. But McKen also collected other plants from around the world and brought them to Durban, setting the foundation for the contemporary gardens. Today the gardens have a global collection of plants, including 6,000 orchids across 75 genera and numerous palms, the oldest of which was planted in 1873. It also has a strong contingent of native plants, perhaps most importantly the *Encephalartos woodii* cycad, which is now extinct in the wild, and the *Ziziphus mucronata*, also known as the buffalo thorn (UmPhafa), the last tree from the area's original forest that once covered what is now the city of Durban.

www.durbanbotanicgardens.org.za +27 (0)31 322 4021

08 KIRSTENBOSCH NATIONAL BOTANICAL GARDEN

Rhodes Drive, Newlands, Cape Town, South Africa

TO VISIT BEFORE YOU DIE BECAUSE

Unlike most botanical gardens, Kirstenbosch is largely dedicated to preserving native species.

When Kirstenbosch National Botanical Garden was founded in 1913 on the slopes of Cape Town's Table Mountain, most botanical gardens had missions to collect and exhibit flora from around the world. But here, the institution's leaders, including its first director, Harold Pearson (1870–1916), simply wanted to preserve South Africa's own plants. Prior to the establishment of the garden, the land it now sits on was used for farming; over the course of 15 years, garden staff labored to clear the land for such features as the Cycad Amphitheatre, Mathews' Rockery, and the Arboretum, through which snakes the garden's iconic Centenary Tree Canopy Walkway, nicknamed "The Boomslang," all of which still exist today. The grounds now cover 1,300 acres and are home to more than 7,000 species of plants.

www.sanbi.org/gardens/kirstenbosch +27 (0)21 799 8783

09 ENTEBBE BOTANICAL GARDEN

12 Kampala-Entebbe Expressway, Entebbe, Uganda

TO VISIT BEFORE YOU DIE BECAUSE

The Milege World Music Festival is hosted in the garden each year.

Set on the shores of Lake Victoria, the Entebbe Botanical Garden opened in 1898 as an agricultural research facility, helping the British establish the tea and cacao industries in the country. Over the following century, the gardens became far more than a farming center; today, hundreds of native and foreign flora fill its grounds, as does wildlife such as birds and monkeys. But Entebbe Botanical Garden's biggest claim to fame—at least if you ask the locals—is that scenes from the Tarzan films of the 1940s were reportedly shot here. Today, the garden is used as a leisure spot for locals and visitors, who visit the garden's beach and café on sunny days. It's also the site of the annual Milege World Music Festival, put on by the Milege Afrojazz Band.

No official website

10 CAMPO DE LOS TULIPANES

Route 259, Km. 50, Trevelin, Argentina

TO VISIT BEFORE YOU DIE BECAUSE

There's nothing like seeing thousands of colorful tulips in bloom beneath the mountains of Patagonia.

Trevelin, Argentina, is a unique place. Set in the Chubut province in the northern reaches of Patagonia, it was settled by Welsh immigrants in the late 19th century, and Welsh culture pervades its community today. Trevelin also has a connection to the Netherlands, not through heritage, but through tulip cultivation; the town is famous for its *campo de los tulipanes*, or tulip fields, which have been grown by the Ledesma family since 1996. Each southern spring—that is, in October and November—thousands of colorful tulips blossom in the fields beneath snow-capped mountains. It makes for quite the sight! The Ledesmas welcome visitors in search of the perfect photograph of their flower fields. Visit at night for some unique photography opportunities.

No official website +54 (0)2945-406335

11 JARDÍN BOTÁNICO CARLOS THAYS

Santa Fe Avenue 3951, Buenos Aires, Argentina

TO VISIT BEFORE YOU DIE BECAUSE

Designed by iconic French landscape architect Carlos Thays, this garden—a national monument—is a piece of living history in Buenos Aires.

After moving to Argentina in 1889, French designer Carlos Thays (1849–1934) quickly became one of the country's foremost landscape architects, developing a number of green spaces across the country, including the botanical garden in Buenos Aires that bears his name. The 17-acre Jardín Botánico Carlos Thays has several geographically themed zones, including a classical French garden and a garden dedicated to the country's indigenous plants, as well as an Art Nouveau-style greenhouse exhibited at the Exposition Universelle of 1889 in Paris. It is also home to a botanical garden in an English-style red-brick mansion where Thays once lived, as well as a butterfly garden, the city's municipal gardening school, and a collection of sculptures. The Argentinian government named the Jardín Botánico Carlos Thays a national monument in 1996.

www.buenosaires.gob.ar/jardinbotanico +54 11 4831-4527

12 JARDÍN JAPONÉS

Casares Avenue 3450, Buenos Aires, Argentina

TO VISIT BEFORE YOU DIE BECAUSE

The Japanese gardens are one of the largest outside of Japan.

At the turn of the 20th century, Argentina's Japanese community saw massive growth, and it eventually developed into the robust group it is today. In Buenos Aires, the friendship between the two countries has been honored by the creation of the Jardín Japonés, managed by the Fundación Cultural Argentino Japonesa. It was developed in 1967 ahead of a visit by Japan's Prince Akihito, who later became emperor, and his wife, Princess Michiko. Today, the 5-acre garden serves not only as a peaceful oasis in the city, but also as a major cultural center, between its living collection of Japanese flora, its restaurant and teahouse, and a visitor center with exhibitions and programming from bonsai workshops to origami classes.

www.jardinjapones.org.ar + 54 11 4804-9141

13 ROSEDAL DE PALERMO

Isabel, Infanta Avenue 900, Buenos Aires, Argentina

TO VISIT BEFORE YOU DIE BECAUSE

There are more than 18,000 roses in this garden, not to mention an Andalusian patio, a Greek bridge, and an amphitheater.

The Rosedal de Palermo, also known as the Paseo del Rosedal, is a delight for all your senses. Located in Buenos Aires' Parque Tres de Febrero, or Bosques de Palermo, the garden has more than 18,000 roses across more than 1,000 varieties—when they're in bloom, they dazzle with their colors and delicate fragrances. The Rosedal was originally conceived by the city's mayor Joaquín Anchorena (1876–1961), then developed by agronomist and landscape designer Benito Javier Carrasco (1877–1958) in 1914. Greek-style architecture was added to the garden in the following years, including a pergola and a bridge over the lake, as well as an Andalusian-style patio, which was donated by the city of Seville, Spain. Be sure to visit the Garden of the Poets, where you'll find 26 busts of bards from around the world.

www.turismo.buenosaires.gob.ar/en/otros-establecimientos/palermo-rose-garden no tel. no. available

14 ANDROMEDA BOTANIC GARDENS

Bathsheba, St. Joseph, Barbados

TO VISIT BEFORE YOU DIE BECAUSE

The tropical gardens are the legacy of acclaimed Barbadian horticulturist Iris Bannochie.

A regular participant at the Chelsea Flower Show in London, self-taught Barbadian scientist Iris Bannochie (1914–1988) traveled the world collecting tropical and subtropical plants for her private collection. Starting in 1954, the horticulturalist would rise at 5 a.m. each day to begin work in her garden. The result is the 6-acre Andromeda Botanic Gardens, kept private until 1976, when Bannochie first opened the property to the public for a fundraising event. Since then, thousands of visitors have been welcomed into the gardens. Bannochie bequeathed Andromeda to the Barbados National Trust upon her death, and today, the organization maintains the garden's hundreds of species of plants—including an impressive collection of palms and a lush lily pond—for Barbadians and visitors alike. It also operates an education program for budding horticulturalists.

www.andromedabarbados.com +1 246-433-9384

15 HUNTE'S GARDENS

Castle Grant, St. Joseph, Barbados

TO VISIT BEFORE YOU DIE BECAUSE

You might just be able to meet Mr. Hunte himself and share a round of rum punches on his veranda.

As you approach Hunte's Gardens, set in a lush gully in Barbados, you'll first hear birds twittering and palm trees rustling, and then the melody of classical music will reach your ears. The garden's maestro, music lover Anthony Hunte, was born into a family of Barbadian horticulturists, so when the opportunity arose for him to purchase a 10-acre plot of land on a former sugar plantation in the early 1990s, he jumped on it, intending to cultivate his own botanical masterpiece. The garden has a theatrical quality to it, with plenty of dramatic flair through colorful blooms and antique sculptures, but it's a lovely place to sit and relax in quiet contemplation. Hunte lives on the property, and if you happen to cross his path, he might just invite you to the veranda for a spirited conversation. Don't miss the chance to sample his rum punch, or to play the 1934 grand piano in the former stable.

www.huntesgardens-barbados.com +1 246-433-3333

16 JARDIM BOTÂNICO DE CURITIBA

Engenheiro Ostoja Roguski Road, Curitiba, Brazil

TO VISIT BEFORE YOU DIE BECAUSE

An Art Nouveau greenhouse, formal French gardens, and an art gallery—the Jardim Botânico de Curitiba has it all.

The Jardim Botânico de Curitiba might have opened only in 1991, but its elements bring visitors to another time and place. The centerpiece of the garden is the Art Nouveau greenhouse, set among formal French gardens. Behind it is an art museum dedicated to Polish-Brazilian artist Frans Krajcberg (1921–2017), a contemporary of Georges Braque (1882–1963), Joan Miró (1893–1983), and Jean Dubuffet (1901–1985), with whom he studied in Paris. Krajcberg is a fitting choice for a botanical garden—a staunch environmentalist, he often used burnt tropical wood from illegal fires in his sculptures to protest deforestation. The Jardim Botânico de Curitiba is also home to a botanical museum that contains an herbarium focusing on native plants.

www.curitiba.pr.gov.br/conteudo/jardim-botanico/287

no tel. no. available

17 JARDIM BOTÂNICO DO RIO DE JANEIRO

Jardim Botânico Road, 1008, Rio de Janeiro, Brazil

TO VISIT BEFORE YOU DIE BECAUSE

A UNESCO biosphere reserve, this massive garden is the flagship of Brazilian horticulture.

Far beneath the right arm of the iconic *Christ the Redeemer* statue in Rio de Janeiro is the city's acclaimed botanical garden, one of the best, if not the best, in Brazil. Created in 1808 by proclamation of the prince regent D. João VI of Portugal (1757–1826), who would eventually become king, to cultivate tea and imported spices like cinnamon and pepper, the Jardim Botânico do Rio de Janeiro has played a large role in Brazilian horticulture. The garden was once home to the very first royal palm tree planted in the country—named Palma Mater—from which all 134 palms in the garden's famous Avenue of Royal Palms descended. Then, under the leadership of Friar Leandro do Sacramento (1778–1829), the garden became a reference institution for Brazilian flora—a role it continues to hold today, with an expansive botanical library. And that's not to mention its living collection, spread across approximately 135 acres of cultivated gardens and 210 acres of natural Atlantic forest. It's no wonder the Jardim Botânico do Rio de Janeiro has been designated a UNESCO biosphere reserve.

www.gov.br/jbrj/pt-br +55 21-3874-1808

18 THE BUTCHART GARDENS

800 Benvenuto Avenue, Brentwood Bay, British Columbia, Canada

TO VISIT BEFORE YOU DIE BECAUSE

These gardens are a National Historic Site of Canada.

After settling in Victoria with her husband at the turn of the 20th century, Jennie Butchart (1866–1950) dreamed of transforming her husband's limestone quarry in their seaside backyard into a world-class garden. While the quarry was still operational, she settled for smaller gardens on their estate, but when the quarry began to run low, it allowed her to more boldly pursue her original ambition. Over the next two decades, the Butcharts turned the former quarry into a paradise filled with themed gardens and lined with colorful flower beds. More than 100 years later, their efforts are still delighting visitors, who enjoy the gardens in every season, including the spectacular Christmas display. In 2004, during its 100th anniversary, The Butchart Gardens was named a National Historic Site of Canada.

www.butchartgardens.com +1 250-652-4422

19 HATLEY PARK GARDENS

2005 Sooke Road, Victoria, British Columbia, Canada

TO VISIT BEFORE YOU DIE BECAUSE

The Edwardian formal gardens are beautifully preserved—and they have an incredible view of the Olympic Mountains.

Canadian industrialist and politician James Dunsmuir (1851–1920) spared no expense when it came to his familial castle in Victoria, British Columbia. "I don't care what it costs, just build it," he was rumored to have said. His 565-acre estate included a Scottish baronial-style castle by architect Samuel Maclure (1860–1929)—you might recognize it from the film *X-Men*—and nine formal gardens designed by landscape architects Franklin Brett and George D. Hall, who studied under Frederick Law Olmsted (1822–1903). The gardens, including an Italian garden thought to be the favorite of Dunsmuir's wife, Laura, and a two-level Japanese garden, remain extremely well preserved to this day as part of the Hatley Park National Historic Site. Visitors can stroll the grounds themselves, taking in the impressive views of the Olympic Mountains, or they can take a guided tour that will provide some historical context.

www.hatleypark.ca/hatley-park-gardens +1 250-391-2666

20 MONTREAL BOTANICAL GARDEN (JARDIN BOTANIQUE DE MONTRÉAL)

4101 Sherbrooke Street East, Montreal, Quebec, Canada

TO VISIT BEFORE YOU DIE BECAUSE

With more than 22,000 species of plants spread across 190 acres, the Montreal Botanical Garden has one of the most extensive collections in the world.

Despite facing the hardships of the Great Depression, Montreal mayor Camillien Houde (1889–1958) established the Montreal Botanical Garden in 1931 at the urging of botanist Brother Marie-Victorin (1885–1944). The institution initially featured only reception gardens around the Art Deco administration building, but in 1956, its exhibition greenhouses with flora from around the world opened to the public, allowing guests to visit year-round. Today, there are more than 30 indoor and outdoor thematic gardens, including a Ming Dynasty–style Chinese garden, designed by landscape architects from the Shanghai Institute of Landscape Design and Architecture, and a First Nations garden dedicated to the Indigenous cultures of Canada. The Montreal Botanical Garden is also home to the Institut de recherche en biologie végétale (IRBV), or the Plant Biology Research Institute, a collaborative project with the Université de Montréal.

www.espacepourlavie.ca/jardin-botanique +1 514-868-3000

21 MUTTART CONSERVATORY

9626 96A Street, Edmonton, Alberta, Canada

TO VISIT
BEFORE YOU DIE
BECAUSE

This is one of Canada's largest indoor gardens—and it's housed inside four glass pyramids.

Unless you're from Alberta, when you picture the most famous pyramids in the world, you're probably not thinking about the four glass pyramids in Edmonton. But for locals, they're an iconic part of the city's skyline. The structures house the Muttart Conservatory, an institution established in 1976 by philanthropists Merrill Muttart (1903–1970) and Dr. Gladys Muttart (1902–1969), founders of the Muttart Foundation. Three of the pyramids are each dedicated to a specific climate—arid, temperate, and tropical—while the fourth hosts a handful of rotating exhibitions a year, often timed to the seasons (think roses in summer and poinsettias in winter). In the center of the four biomes, you'll find the Centre Court, where artists Keith Walker and Alex Janvier have works installed.

www.edmonton.ca/attractions_events/muttart-conservatory

+1 780-442-5311

22 VIÑA SANTA RITA

Camino Padre Hurtado 0695, Alto Jahuel, Buin, Santiago, Chile

TO VISIT BEFORE YOU DIE BECAUSE

You get to combine your garden visit with a wine tasting at one of Chile's oldest wineries.

It's not all about vino at the Viña Santa Rita, a winery about 45 minutes outside of Santiago, Chile. While there are nearly 2,500 acres of vineyards on the estate, there's also a 99-acre park filled with European-style gardens and a 19th-century manor house, the former private home of Santa Rita's founder, which now houses the Casa Real Hotel. The gardens were designed by French landscape architect Guillaume Renner (1843–1924) and blend English, French, and Italian styles; you'll find a combination of long walks, old groves, flower beds, hedge mazes, fountains, and even Roman-style baths here. Visitors to the estate can book tours that include the gardens—and, of course, a wine tasting.

www.santarita.com (+562) 2362 2000

23 JARDÍN BOTÁNICO JOAQUÍN ANTONIO URIBE

Calle 73 #51D 14, Medellín, Colombia

TO VISIT BEFORE YOU DIE BECAUSE

The Orquideorama, dedicated to Colombia's national flower, is a futuristic architectural masterpiece.

When Medellín was selected to host the 1972 World Conference of Orchidology, the city's gardening groups joined together to establish what would become the Jardín Botánico Joaquín Antonio Uribe, named for the famed Colombian naturalist and writer. The institution was the main venue for the event, which saw 62 delegations from 16 countries meet to discuss the orchid, Colombia's national flower. But in subsequent decades, the city faced difficult times, and the garden languished. In 2005, however, Medellín mayor Sergio Fajardo (b. 1956) began a largely successful urban renewal project that included a renovation of the garden—one of the new additions was the Orquideorama, a honeycomb-like structure by Plan B Architects and JPRCR Architects designed to house the exhibit of orchids, while another was In Situ, one of Medellín's best restaurants.

www.botanicomedellin.org + 57 4 444 55 00

24 TULCÁN MUNICIPAL CEMETERY

Cotopaxi Avenue and El Cementerio Avenue, Tulcán, Ecuador

TO VISIT
BEFORE YOU DIE
BECAUSE

Where else can you find a cemetery decorated with topiaries depicting totems from ancient cultures?

When a 1923 earthquake damaged the old municipal cemetery in Tulcán, Ecuador, a replacement was built in 1936. Its caretaker, gardener José María Azael Franco (1900–1985), decided to spice things up a bit at the new site. As it turns out, Franco was a skilled topiary artist, and he shaped the cypress trees throughout the graveyard into totems from world cultures, from the Inca to the Augustinian monks. The government of Ecuador was so impressed with the work, it honored the cemetery as a national cultural site in 1985. A year later, Franco passed away, and he was appropriately buried in the cemetery. His epitaph famously reads "In Tulcán, a cemetery so beautiful that it invites one to die!" Today, Franco's sons maintain the topiaries, while locals honor the landmark by carving topiaries inspired by the originals all throughout town.

No official website

25 JARDÍN BOTÁNICO DE QUITO

Pasaje #34, Rumipamba E6-264 and De Los Shyris Avenue, Quito, Ecuador

TO VISIT BEFORE YOU DIE BECAUSE

The garden's orchid house has an impressive selection of the more than 4,300 species native to Ecuador.

In 1989, the Museo Ecuatoriano de Ciencias Naturales (the Ecuadorian Museum of Natural Sciences), the Club de Jardinería de Quito (the Quito Gardening Club), and the Quito government jointly agreed to develop a botanical garden in the city. Two years later, the Fundación Botánica de los Andes (FBA), or the Botanical Foundation of the Andes, was created to preserve the flora of the Ecuadorian Andes. The Jardín Botánico de Quito itself, however, took some time to build, opening to the public only in 2005. While its primary focus is on the ecosystems of Ecuador, the garden has specimens from around the world. Some of the garden's highlights include an orchid house with nearly 2,000 orchids, representing a portion of the country's 4,300 species as well as international ones, a greenhouse of carnivorous plants, and a bonsai museum.

www.jardinbotanicoquito.com/es +593 998379482

26 FLOATING GARDENS OF XOCHIMILCO

Xochimilco, Mexico City, Mexico

TO VISIT
BEFORE YOU DIE
BECAUSE

The well-preserved "floating" islands and ancient canal system are Aztec agriculture and transportation marvels. Come for the islands; stay for the party.

Take a glimpse into Mexico's pre-Hispanic history at the Floating Gardens of Xochimilco, a UNESCO World Heritage site some 17 miles outside the heart of Mexico City. To make use of the shallow lake, the Aztecs developed a system of artificial "floating" islands called *chinampas* on which they could grow crops. Today, some *chinampas* still exist on Lake Xochimilco, but the real reason visitors flock to the region is to ride the *trajineras*, colorful gondola-like boats, which locals often do for celebratory events or a fun (and perhaps boozy) afternoon with family and friends. Hire a captain, buy some street food from the floating vendors, and set sail on the ancient canals.

No official website

27 JARDÍN BOTÁNICO DEL INSTITUTO DE BIOLOGÍA DE LA UNAM

Tercer Circuito exterior (no street number),
Ciudad Universitaria Coyoacán, Mexico City, Mexico

TO VISIT BEFORE YOU DIE BECAUSE

It'll help you understand how truly biodiverse Mexico is.

Many tourists come to Mexico for the historic sites, the beaches, and the food. But you might want to add flora to your must-see list in the country. Mexico has an incredibly diverse range of plant life—in fact, it has more plant species than the United States and Canada combined. Get a good primer on the country's biodiversity at the botanical garden of the Universidad Nacional Autónoma de México (UNAM). The second-oldest botanical garden in the country, it has an impressive collection of flora, including a number of Mexico's 800-plus endemic species of cacti, many of which are endangered. Plants aside, the garden is known for its intriguing rock formations; it's set on an old lava flow, which has developed natural waterfalls, grottoes, and rockeries over time. The garden is also home to the rare Tarántula del Pedregal spider (*Aphonopelma anitahoffmannae*), which lives only in protected pockets of Mexico City.

www.ib.unam.mx/jardin + 52 56 22 90 63

28 LAS POZAS

Camino Paseo Las Pozas (no street number), Xilitla, San Luis Potosí, Mexico

TO VISIT BEFORE YOU DIE BECAUSE

The gardens are a Surrealist dreamscape by artist and poet Edward James.

Born in Scotland to an upper-class family, poet Edward James (1907–1984) built a name for himself as a fervent supporter of the Surrealist art movement, sponsoring artist Salvador Dalí (1904–1989), who famously referred to his patron as "crazier than all the Surrealists together," building a collection of Surrealist masterpieces, and even posing for two portraits by René Magritte (1898–1967), including *Not to Be Reproduced*. After moving to North America, James set out to build a personal Garden of Eden, ultimately purchasing a remote site in the Mexican jungle. There, over the course of 40 years, he filled his property with concrete sculptures and structures, as if the impossible architecture of Dalí's paintings had sprung to life. Visiting Las Pozas might be a bit of a journey—it's a four-and-a-half-hour drive from San Luis Potosí and a seven-and-a-half-hour drive from Mexico City—but the payoff is worth it.

www.laspozasxilitla.org.mx

29 DIAMOND BOTANICAL GARDENS

Soufrière Estate, Diamond Road, Soufrière, St. Lucia

TO VISIT
BEFORE YOU DIE
BECAUSE

You can enjoy tropical flora, a waterfall, and a mineral bath all in one place.

In 1713, King Louis XVI (1754–1793) granted the three Devaux brothers a parcel of land on the Caribbean island of St. Lucia. The trio would move to the island in 1740 to develop their property, which today is the grand Soufrière Estate, home of the Diamond Botanical Gardens. Though originally a lime, cacao, and copra plantation, the historic estate had a second calling: a health destination known for its restorative mineral baths. (After samples from the property's springs were sent to France for analysis in 1784, it was determined that Soufrière's mineral waters had the same quality as those found in the spa towns of Aix-les-Bains and Aix-la-Chapelle (Aachen) in France and Germany, respectively.) Formal baths were installed on the site, though they fell into disrepair until the 1930s, when André du Boulay (1898–1982) restored them for his private use. Under the guidance of Joan Devaux (b. 1922), du Boulay's daughter, the estate was transformed into a tourist destination featuring 6 acres of tropical gardens, the iconic Diamond Falls, and public mineral baths for soaking.

www.diamondstlucia.com +1 758-459-7155

30 ASTICOU AZALEA GARDEN

3 Sound Drive, Northeast Harbor, Maine, United States of America

TO VISIT BEFORE YOU DIE BECAUSE

It's the perfect stop on a trip to Acadia National Park.

At just over 2 acres, the Asticou Azalea Garden in coastal Maine is much smaller than nearby Acadia National Park, allowing you to interact with nature in an entirely different way. The garden was planted in 1957 by Charles K. Savage (1903–1979), a local innkeeper, who sought to blend the landscape of Mount Desert Island with design elements from classical Japanese gardens. In the spring, a cherry tree blossoms, while azaleas and rhododendrons usher in summer. Later in the season, the water lilies bloom, and the fall dazzles with the deciduous trees' traditional colors of the season. There's also a sand garden here designed to mimic the 16th-century one at Ryōan-ji in Kyoto, Japan.

www.gardenpreserve.org/asticou-azalea-garden

+1 207-276-3727

31 BALBOA PARK

1549 El Prado, San Diego, California, United States of America

TO VISIT BEFORE YOU DIE BECAUSE

With 19 gardens across 1,200 acres of land, Balboa Park is an urban horticulturalist's dream.

Balboa Park is San Diego's crown jewel of green space, with more than 1,200 acres of grounds that are home to 19 different gardens (not to mention 17 museums and cultural institutions!). Originally established as "City Park" in 1868 by San Diego's urban planners, then renamed in 1910 for Spanish explorer Vasco Núñez de Balboa, Balboa Park received its first plantings in 1892 at the hands of botanist and landscape architect Kate Sessions (1857–1940), warmly regarded as "The Mother of Balboa Park." Over the park's 100-plus-year history, its collection of gardens has grown to include the Alcazar Garden, modeled after the gardens of the Alcázar of Seville, Spain; the Japanese Friendship Garden, which honors San Diego's sister city of Yokohama, Japan; and the 3-acre Inez Grant Parker Memorial Rose Garden, one of the world's foremost rose gardens.

www.balboapark.org/explore/gardens +1 619-239-0512

32 BROOKLYN BOTANIC GARDEN

990 Washington Avenue, Brooklyn, New York, United States of America

TO VISIT BEFORE YOU DIE BECAUSE

Its cherry blossom festivities and bonsai museum are not to be missed.

As New York's urban sprawl spread throughout the boroughs in the late 19th century, city leaders knew they would have to preserve green space. In 1897, 39 acres in Brooklyn were set aside for the creation of a botanic garden. The Brooklyn Botanic Garden (BBG) officially opened in 1911, growing to 52 acres over the next century. More than just a garden, BBG is also a cultural center, with a particular focus on Japan. Its Japanese Hill-and-Pond Garden, designed by landscape architect Takeo Shiota (1881–1943), was one of the first Japanese gardens to be built in the United States when it was completed in 1915. In 1947, the garden hired its first bonsai curator, Frank Okamura (1911–2006), and today is the garden home to the C.V. Starr Bonsai Museum. In 1980, the city of Tokyo gifted the BBG a 500-year-old shogun lantern, and in 1982, the garden hosted its first Sakura Matsuri cherry blossom festival—there are several dozen species of cherry blossom trees in the garden that put on a spectacular show each spring.

www.bbg.org +1 718-623-7200

33 CONSERVATORY OF FLOWERS

100 John F. Kennedy Drive, San Francisco, California, United States of America

TO VISIT BEFORE YOU DIE BECAUSE

The Victorian-style wood-and-glass conservatory is the oldest in North America.

Real estate magnate and horticulturalist James Lick (1796–1876), once the wealthiest man in California, had a vision for two grand Victorian-style conservatories to be built at his home in Santa Clara. Unfortunately, he died before they were realized. The building materials, however, had already been sent to California, and ultimately were gifted to the city of San Francisco. While one conservatory was never built, the other was erected in Golden Gate Park in 1879 by greenhouse manufacturers Lord & Burnham and today is the home of the Conservatory of Flowers. As you can infer from its name, the conservatory houses ... well, flowers! The collection covers different ecosystems across five galleries, highlighting rare and endangered species from diverse destinations like Ecuador's cloud forests and Sumatra's rainforests.

www.conservatoryofflowers.org +1 415-831-2090

34 DESERT BOTANICAL GARDEN

1201 N. Galvin Parkway, Phoenix, Arizona, United States of America

TO VISIT BEFORE YOU DIE BECAUSE

It's one of the few horticultural institutions dedicated to the harsh desert climate.

You might think that the desert is a hostile environment—and it is. But it's not one devoid of plant life. The Desert Botanical Garden in Phoenix, Arizona, explores the diversity of arid-climate flora, from cacti to wildflowers. Founded in 1939 by local plant lovers, including Swedish botanist Gustaf Starck (1871–1945) and socialite Gertrude Divine Webster (1872–1947), the garden has grown to 150 acres with a collection of more than 50,000 desert plants today. Explore five thematic hiking paths to peruse the plant life, learning about topics like the relationship between desert plants and the Sonoran Desert's Indigenous cultures. The Desert Botanical Garden also hosts a Desert Landscape School to educate both casual and professional gardeners about how to harmoniously work with the environment around them.

www.dbg.org +1 480-941-1225

35 FAIRCHILD TROPICAL BOTANIC GARDEN

10901 Old Cutler Road, Miami, Florida, United States of America

TO VISIT BEFORE YOU DIE BECAUSE

The garden was developed by someone who could be considered botanic royalty: Dr. David Fairchild.

Before retiring in Miami in 1935, plant explorer Dr. David Fairchild (1869–1954) traveled the world for nearly four decades, collecting plants that had potential uses for the American people, from produce like mangoes to the flowering cherry blossom trees planted throughout Washington, D.C. But his retirement didn't last long. With the help of former accountant Col. Robert H. Montgomery (1872–1953), Dr. Fairchild established the 83-acre Fairchild Tropical Botanic Garden in 1938, which featured grounds by acclaimed Floridian landscape architect William Lyman Phillips, a colleague of Frederick Law Olmsted. In 1940, he took his first collecting exhibition—to Southeast Asia. Over the decades, the garden's facilities and collections have grown greatly—it even has a 2-acre tropical rainforest grown from Amazonian plants. Today, the Fairchild Tropical Botanic Garden is a leading institution for the protection of endangered tropical plants.

www.fairchildgarden.org +1 305-667-1651

36 FORT WORTH BOTANIC GARDEN

3220 Botanic Garden Boulevard, Fort Worth, Texas, United States of America

TO VISIT BEFORE YOU DIE BECAUSE

The historic botanical garden is one of the best examples of municipal rose gardens in the country.

Part of the Botanical Research Institute of Texas, the Fort Worth Botanic Garden (FWBG) is the oldest major botanic garden in the state of Texas, established in 1934. But the garden's full history predates that opening. In 1912, the Fort Worth park board purchased 33 acres of land that would become Rock Springs Park. Over a quarter century, landscape architecture firm Hare & Hare would refine the park, developing what is now one of the finest surviving examples of American municipal rose gardens; the ramps and terraces were inspired by Italy's Villa Lante, while the water cascade was inspired by France's Versailles. Eventually, the park would become the FWBG, which has since expanded to 109 acres, 25 specialty gardens (including the popular Japanese Garden), and 2,500 plant species.

www.brit.org +1 817-463-4160

37 HO'OMALUHIA BOTANICAL GARDEN

45–680 Luluku Road, Kāne'ohe, Oahu, Hawaii, United States of America

TO VISIT BEFORE YOU DIE BECAUSE

It has one of the most impressive mountain backdrops of all time.

One of the five Honolulu Botanical Gardens, Ho'omaluhia is a quiet 400-acre area in the shadow of the looming Ko'olau Mountains. Built by the U.S. Army Corps of Engineers to protect Kāne'ohe from floods, the garden is a peaceful refuge—in fact, that's the literal translation of its name—filled with plantings of tropical flora from regions across the world. After you've finished marveling at the plants and the ridges behind the garden on the walking trails, you can fish in the 32-acre lake, take a drawing or painting class, or even camp overnight in the verdant rainforest. (An aside: Please don't stop on the entrance road to snap a photo! There are safer and equally impressive views throughout the park—like from the Kilonani Mauka Overlook, the highest point in the garden.)

www.honolulu.gov/parks/hbg.html?id=569:ho +1 808-233-7323

38 LADEW TOPIARY GARDENS

3535 Jarrettsville Pike, Monkton, Maryland, United States of America

TO VISIT BEFORE YOU DIE BECAUSE

Harvey Ladew created more than 100 impressive topiaries—without any formal training.

Lured to the Maryland countryside by the promise of fox hunts, Harvey S. Ladew (1887–1976) purchased more than 200 acres of farmland for his new domicile in 1929. In addition to renovating and expanding the old farmhouse on the property, Ladew also planted 22 acres of formal gardens on his own, without any professional training. "When I first bought the farm, I decided to do all the landscaping and planning of the gardens myself, although I knew I would make a lot of mistakes," he once wrote. "I made the mistakes, but they taught me the little I know about gardening." Impressively, Ladew not only developed beautiful "garden rooms" dedicated to roses, pink flowers, and yellow blooms, but also became a masterful topiary artist, adding more than 100 living sculptures to his property—including, of course, a fox hunt scene. His efforts did not go unnoticed. Ladew received the Garden Club of America's Distinguished Achievement Award for his magnificent topiary garden.

www.ladewgardens.com +1 410-557-9466

39 LEWIS GINTER BOTANICAL GARDEN

1800 Lakeside Avenue, Richmond, Virginia, United States of America

TO VISIT BEFORE YOU DIE BECAUSE

This quintessential botanical garden has it all: themed gardens, a historic home, a classic conservatory, and a beautiful holiday light show.

Though the land that the Lewis Ginter Botanical Garden sits on has a long history—they were once the Oughnum hunting grounds of the Powhatan people, then the private property of Virginia governor Patrick Henry, and eventually the site of the Lakeside Wheel Club bicycling organization—the gardens themselves are relatively new. After the founder of the Lakeside Wheel Club, local businessman Lewis Ginter (1824–1897) passed away, his niece, philanthropist Grace Arents (1848–1926), purchased the property, eventually willing it to the City of Richmond with the stipulation that it be transformed into a botanical garden in honor of her uncle. Thus, the Lewis Ginter Botanical Garden was born in 1984, and today it covers 89 acres with more than 5,700 taxa of plants. Don't miss its annual GardenFest of Lights holiday show.

www.lewisginter.org +1 804-262-9887

40 LIMAHULI GARDEN AND PRESERVE

5–8291 Kuhio Highway, Hanalei, Kauai, Hawaii, United States of America

TO VISIT BEFORE YOU DIE BECAUSE

The site is one of the most biodiverse places in Hawaii—and a culturally significant one, too.

There's a good reason the Hawaiian island of Kauai is known as the "Garden Island." It's one of the greenest places on Earth, covered head to toe (or mountain to valley) in rainforests. That's why it's not altogether surprising that four of the five National Tropical Botanical Garden properties are located here, including the scenic Limahuli Garden and Preserve. (The fifth garden is located in Florida.) Limahuli is not only a garden set in one of the most biodiverse locations in the state, home to a variety of native and Polynesian-introduced flora, but also an archaeological site; the land was once part of the *ahupuaʻa*, or traditional community, of Hāʻena. During your visit, you can learn about the ancient Hawaiians' relationship to the land through the garden's mission of biocultural conservation.

www.ntbg.org/gardens/limahuli +1 808-826-1053

41 LONGWOOD GARDENS

1001 Longwood Road, Kennett Square, Pennsylvania, United States of America

TO VISIT BEFORE YOU DIE BECAUSE

One of the great American gardens, Longwood draws in more than 1.5 million visitors a year to its 1,100 acres of plantings.

Once Native American land, then Quaker farmland in the 18th century, the property that is now known as Longwood Gardens grew its horticultural roots in 1798, when brothers Samuel Peirce (1766–1838) and Joshua Peirce (1766–1851) planted a 15-acre arboretum on their great-grandfather's property. That arboretum still exists at Longwood today as Peirce's Park—but it's far from the only attraction there. Under the vision of Pierre du Pont (1870–1954), who purchased the Peirce estate in 1906, Longwood has grown into an astonishing complex of indoor and outdoor gardens intermingled with historic architecture, from conservatories to fountains. It also has one of the country's best holiday light displays, as well as year-round cultural and artistic programming.

www.longwoodgardens.org +1 610-388-1000

42 MISSOURI BOTANICAL GARDEN

4344 Shaw Boulevard, St. Louis, Missouri, United States of America

TO VISIT BEFORE YOU DIE BECAUSE

If the vast collection of rare and endangered plants isn't enough, the garden has an impressive architectural heritage, too.

Though Henry Shaw (1800–1889) was a successful businessman, his real interest was botany, which is why he founded the Missouri Botanical Garden in St. Louis in 1859. Continuously in operation since then, the garden is renowned for its collection of 51,000 live plants (including a number of rare and endangered species that its conservation team is working to recover), its 6.5-million-specimen herbarium, and a slew of architecturally significant buildings. On the last front, the 79-acre campus is home to Shaw's 1850s country home, a Russian-inspired observation tower, and the Climatron, the world's first geodesic dome greenhouse, among numerous other structures. The Missouri Botanical Garden also serves as a cultural center, putting on annual events like Chinese Culture Days and the Japanese Festival, the Whitaker Music Festival, and the Grapes in the Garden wine-tasting program.

www.missouribotanicalgarden.org +1 314-577-5100

43 MYRIAD BOTANICAL GARDENS

301 W. Reno, Oklahoma City, Oklahoma,
United States of America

TO VISIT BEFORE YOU DIE BECAUSE

There's always something fun happening here, whether it's a theater performance in the amphitheater, a flower festival, floral arrangement classes, or even a children's tea party.

In 1964, acclaimed architect I. M. Pei (1917–2019) was tasked with the grand revitalization of downtown Oklahoma City. While only a portion of his forward-thinking urban plan was realized, one of the greatest accomplishments to come out of it was the Myriad Botanical Gardens, which Pei modeled after Copenhagen's Tivoli Gardens. The centerpiece of the gardens is the Crystal Bridge Tropical Conservatory. Forget the classic Art Nouveau–style greenhouse—this is a monumental concrete, steel, and acrylic cylinder suspended over a sunken koi lake (the term "pond" wouldn't do it justice). Surrounding this are thousands of plants from around the world, an amphitheater, a children's play area with a splash park and merry-go-round, an ice-skating rink, and a restaurant. As you stroll throughout, keep an eye out for wildlife like rabbits, turtles, and waterfowl.

www.oklahomacitybotanicalgardens.com +1 405-445-7080

44 NEW YORK BOTANICAL GARDEN

2900 Southern Boulevard, Bronx, New York, United States of America

TO VISIT BEFORE YOU DIE BECAUSE

The United States' largest botanical garden has more than a million plants, plus a dazzling annual orchid show and blockbuster art exhibitions.

It's hard to believe that ever-crowded New York City is home to the United States' largest botanical garden, but thanks to botanists Nathaniel Lord Britton (1859–1934) and his wife, Elizabeth (1858–1934), it's a true statement. Spread out across 250 acres in the borough of the Bronx, the New York Botanical Garden has a collection of more than a million plants, including 30,000 trees. Its herbarium is the second largest in the world, with 7.8 million specimens. If that weren't enough, it also has a significant research institution that employs more than 100 scientists working on global projects. But back to the gardens themselves—there are more than 50 specialty gardens, including the Victorian-style Enid A. Haupt Conservatory and the award-winning Peggy Rockefeller Rose Garden. Each year, the New York Botanical Garden puts on a thematic orchid show with remarkable floral installations, as well as a major visual art exhibition that follows a botanical theme. Past artists have included Yayoi Kusama (b. 1929), Roberto Burle Marx (1909–1994), Georgia O'Keeffe (1887–1986), and Dale Chihuly (b. 1941).

www.nybg.org +1 718-817-8700

45 PHILADELPHIA'S MAGIC GARDENS

1020 South Street, Philadelphia, Pennsylvania, United States of America

TO VISIT BEFORE YOU DIE BECAUSE

This eclectic art garden is made from thousands of colorful mosaic pieces.

When artist Isaiah Zagar (b. 1939) moved to Philadelphia's South Street with his wife, Julia, in 1968, it was far from the trendy district it is today. To help beautify the neighborhood, the Zagars and other local artists began installing art throughout the streets. For Zagar, that meant colorful mosaic murals. He discovered the folk art of Clarence Schmidt (1897–1978) as a teenager, then went on to study painting at the Pratt Institute in New York before traveling the world to learn from global folk art communities. While decorating the streets of South Philadelphia, he embarked upon his biggest work yet: transforming a vacant lot into an immersive mosaic and found-object installation that now spans indoor galleries and a sculpture garden. Though at one point slated for demolition, the project has been preserved as Philadelphia's Magic Gardens.

www.phillymagicgardens.org +1 215-733-0390

UNCONFINED

46 PORTLAND JAPANESE GARDEN

611 SW Kingston Avenue, Portland, Oregon, United States of America

TO VISIT BEFORE YOU DIE BECAUSE

Former Japanese ambassador Nobuo Matsunaga declared the garden "the most beautiful and authentic Japanese garden in the world outside of Japan."

Gardens have a magic about them that transports you anywhere in the world through the landscape. This is doubly true of Portland Japanese Garden in the city's Washington Park, which has been praised by former Japanese ambassador Nobuo Matsunaga (1923–2011) as "the most beautiful and authentic Japanese garden in the world outside of Japan." Designed by Professor Takuma Tono (1891–1985) of Tokyo Agricultural University, the garden was designed to facilitate healing, cultural exchange, and understanding between the United States and Japan in the postwar years. Ultimately, its mission was and still is to share the "art of craft, connection to nature, experience of peace" with the world, which it certainly achieves through its eight styles of gardens spread across 12 wooded acres. Portland Japanese Garden is also a hub of Japanese architecture, from a traditional teahouse built in Japan and shipped to Portland, to the Cultural Village designed by contemporary Japanese architect Kengo Kuma (b. 1954) in 2015.

www.japanesegarden.org +1 503-223-1321

47 SAN FRANCISCO BOTANICAL GARDEN

1199 9th Avenue, San Francisco, California, United States of America

TO VISIT BEFORE YOU DIE BECAUSE

The garden is home to a grove of hundred-year-old redwood trees, right in the middle of San Francisco.

Set inside Golden Gate Park, the San Francisco Botanical Garden (SFBG) was first dreamed up by the park's supervisor John McLaren (1846–1943) in the late 19th century, but it wasn't funded until 1926, thanks to the bequest of Helene Strybing (1845–1926). It would be another 14 years until the garden officially opened as Strybing Arboretum; its first director, Eric Walther (1892–1959), guided its construction with the help of the Works Progress Administration (WPA), creating a small garden with a collection of global plants. The SFBG has been growing ever since, with the development of a new master plan by landscape architect Robert Tetlow (1922–1988) in 1959, followed by the establishing of the Helen Crocker Russell Library of Horticulture in 1972, and the addition of the world's first Southeast Asian Cloud Forest garden in the early 2000s, among other projects. Today the garden covers 55 acres and features some 9,000 plants, including hundred-year-old redwood trees, a bamboo forest, and a significant collection of magnolias.

www.sfbg.org +1 415-661-1316

48 UNITED STATES BOTANIC GARDEN

100 Maryland Avenue SW, Washington, D.C., United States of America

TO VISIT BEFORE YOU DIE BECAUSE

This is the oldest continuously operating botanic garden in the United States.

When founding fathers George Washington (1732–1799), Thomas Jefferson (1743–1826), and James Madison (1751–1836) conceived of the new nation's capital city at the end of the 18th century, they imagined a national botanic garden for it, too. In 1820, that vision was realized: Madison, then president, formally designated land in Washington, D.C., as the site for the garden. Though that original facility went defunct in 1837, a new garden was built to house plants brought back to the country by the United States Exploring Expedition of 1838–1842. Since 1850, the Victorian-style conservatory of the United States Botanic Garden has been open to the public, making the garden the oldest continuously operating botanic garden in the country. Today, some highlights include the odiferous corpse flower, the elaborate Bartholdi Fountain by French sculptor Frédéric Auguste Bartholdi (1834–1904), and three original plants from the expedition that led to the second founding of the garden (a queen sago, a blue cycad, and a vessel fern).

www.usbg.gov +1 202-225-8333

49 VIZCAYA MUSEUM AND GARDENS

3251 South Miami Avenue, Miami, Florida, United States of America

TO VISIT BEFORE YOU DIE BECAUSE

The Italian Renaissance-style gardens are nothing short of a masterpiece.

In the United States, most formal gardens are modeled after the English or French style. At Vizcaya, it's all about Italian Renaissance splendor. They're the work of industrialist James Deering (1859–1925), who built his grand estate along Biscayne Bay in Miami's Coconut Grove between 1914 and 1922. His formal gardens were designed by Diego Suarez (1888–1974), a Colombian-Italian landscape architect who studied at Accademia di Belle Arti in Florence, where he first met Deering to give him a tour of great Italian gardens. It's not at all surprising, then, that the pair was largely inspired by the experience to create the formal gardens at Vizcaya. Though hurricanes have changed the gardens over the years—the Great Hurricane of 1926, for instance, killed all the plantings in the magnificent rose garden, which wasn't restored until the current day—the gardens remain as impressive as ever. Beyond the 10-acre formal gardens, which include a vast orchid collection with some 2,000 specimens, there are 40 acres of native forests, from rockland hammocks to mangroves.

www.vizcaya.org +1 305-250-9133

50 LALBAGH FORT

Dhaka, Bangladesh

TO VISIT BEFORE YOU DIE BECAUSE

Though unfinished, Lalbagh Fort and its gardens are impressive nonetheless—and they might be haunted.

Prince Muhammad Azam (1653–1707) of the Mughal Empire commissioned the construction of Lalbagh Fort in Dhaka in 1677, during his tenure as viceroy of Bengal. But when he left Delhi, he passed management of the project to his successor, Shaista Khan (c. 1600–1694). Several years into construction, however, Khan's daughter Pari Bibi (?–1684) died prematurely, and fearing that Lalbagh Fort was cursed, he abandoned it. Only three buildings were completed: a mosque, an audience hall, and the mausoleum of Pari Bibi (yes, she's buried there). But lucky for Dhaka's residents today, the foundation for the Persian-inspired gardens was laid before construction ended, and today, they serve as a park and a tourist attraction.

No official website

51 HUMBLE ADMINISTRATOR'S GARDEN

178 Dongbei Street, Gusu District, Suzhou, China

TO VISIT BEFORE YOU DIE BECAUSE

This is a quintessential classical Chinese garden, and perhaps the finest example in the country.

If there's one thing the ancient city of Suzhou, China, is known for, it's its gardens. The first were laid in the sixth century BCE as royal hunting gardens, but over the centuries, private gardens took root. Today more than 50 such gardens still exist in Suzhou, of which nine, including the Humble Administrator's Garden, have been jointly designated a UNESCO World Heritage site. This formerly private garden was built in the early 16th century during the Ming Dynasty as a private retreat for an imperial administrator, who was inspired by a poem to seek a humble life in gardening—hence the site's name. The garden, however, might not be so humble after all. It's the city's largest at nearly 13 acres, with 48 classical Chinese buildings surrounding a lake. It's one of the finest examples of classical Chinese gardens, an art form designed to replicate the landscape in miniature, expressing harmony between humans and nature.

www.szzzy.cn +86 512-962015

52 MASTER OF THE NETS GARDEN

11 Kuojiatou Alley, Daichengqiao Road, Gusu District, Suzhou, China

TO VISIT BEFORE YOU DIE BECAUSE

It's small in scale, but big in impact.

At just over an acre in size, the Master of the Nets Garden may be petite by Suzhou standards, but it's no less spectacular than its larger peers. One of the nine UNESCO-designated gardens in the Chinese city, the garden was first planted in the 12th century by a Song-era government official inspired by the simple life of a fisherman. Though it fell into disrepair as time passed, a Qing-era scholar purchased the estate in the 18th century, restored it, and renamed it Master of the Nets Garden, in honor of its original inspiration. As with all classical Chinese gardens, this one blends the human-made and the natural, incorporating pavilions, trees, and water features to create a harmonic scene. The Astor Chinese Garden Court at the Metropolitan Museum of Art in New York, the first permanent cultural exchange between China and the United States, is based on the design of the Master of the Nets Garden.

www.szwsy.com

no tel. no. available

53 YUANMINGYUAN PARK

Qinghuan West Road, Haidian District, Beijing, China

TO VISIT BEFORE YOU DIE BECAUSE

A foil to the nearby Summer Palace, these ruins showcase a lost piece of Chinese cultural history.

The Chinese name of this park, Yuanmingyuan, translates to "The Garden of Perfect Brightness," but you won't necessarily find an idyll here. Instead, you'll be confronted with the ruins of one of the greatest palace complexes and imperial gardens in China. Built in the 18th century as a summer retreat for Qing emperors, the Old Summer Palace has been compared to France's Versailles—it even had European-style palaces, in addition to traditional Chinese ones. But the palaces and their 800 acres of gardens were looted by Anglo-French forces during the Second Opium War, then burned at the order of Lord Elgin (1811–1863) in response to the killing of British envoys to China. Now, ruins dot the former gardens, which have since been transformed into a public park. Visit the site before you head to the nearby "new" Summer Palace, which was also destroyed during the conflict but has since been rebuilt.

www.yuanmingyuanpark.cn

54 SUMMER PALACE

19 Xinjiangongmen Road, Haidian District, Beijing, China

TO VISIT BEFORE YOU DIE BECAUSE

While classical Chinese gardens are usually small-scale versions of nature, there's nothing miniature about this imperial complex.

After Anglo-French forces razed the original 18th-century imperial palace complex on Longevity Hill during the Second Opium War, it was rebuilt as a summer palace for Empress Dowager Cixi in the late 19th century (unlike the nearby Old Summer Palace, which remains in ruins). Although it suffered damage during the Boxer Rebellion of 1900, it has since been restored and opened to the public as both a historic site with hundreds of buildings and a 717-acre public garden hailed by UNESCO, which named the Summer Palace a World Heritage site in 1998, as "the culmination of several hundred years of Imperial garden design." You could easily spend hours meandering the grounds alone, but don't miss taking in the views of the landscape from a boat ride on Kunming Lake.

www.summerpalace-china.com +86 (0)10 6288 1144

55 WANGJIANGLOU PARK

30 Wangjiang Road, Wuhou District, Chengdu, China

TO VISIT BEFORE YOU DIE BECAUSE

It's a bamboo Garden of Eden.

Chengdu, China, is best known for its Chengdu Research Base of Giant Panda Breeding, but if you want to get an up-close look at the bears' favorite food, head to Wangjianglou. Dedicated to the Tang-era poet Xue Tao (768–831), the park is filled with more than 200 species of towering bamboo, one of her favorite plants. After you amble through Wangjianglou's robust bamboo gardens, you can climb the Wangjiang Tower, also known as the Chongli Pavilion, to get a bird's-eye view of the gardens, then visit the cultural pavilions to learn more about Xue Tao and to see historic artifacts. Before you leave, stop at one of the teahouses that line the Fu River.

No official website

56 YU GARDEN

279 Yuyuan Old Street, Huangpu District, Shanghai, China

TO VISIT BEFORE YOU DIE BECAUSE

This is Shanghai's most famous classical garden.

Cosmopolitan Shanghai might be a modern, global city, but behind the futuristic skyscrapers of Pudong district and the Art Deco building of the Bund are nods to China's dynastic past. The Yu Garden in Old Town, also known as Yuyuan, is one such example. Built in 1559, the gardens showcase a little bit of everything when it comes to classical Chinese landscape design; there are rockeries, koi ponds, Ming Dynasty pavilions, and arched bridges enclosed in a 5-acre area protected by a famous dragon wall. And don't miss the 5-ton Exquisite Jade Rock (Yu Ling Long), a craggy boulder used to represent mountains—classical Chinese gardens are all about replicating nature in miniature, after all. Yu Garden is certainly beautiful, but it's not one of the more tranquil gardens. As one of the most popular tourist destinations in Shanghai, it is often filled with crowds.

No official website

+86 (0)21 6326 0830

57 NAN LIAN GARDEN

60 Fung Tak Road, Diamond Hill, Kowloon, Hong Kong

TO VISIT BEFORE YOU DIE BECAUSE

It's a scenic escape from the hubbub of Hong Kong.

A collaboration between the Chi Lin Nunnery and the municipality of Hong Kong that was completed in 2006, the Nan Lian Garden provides a quiet, contemplative space for visitors to enjoy the classic Tang-style landscaping. Follow the meandering path through the garden that covers nearly 9 acres, past koi and lotus ponds, waterfalls, rock gardens, arched bridges, and a gilded pagoda. There's also a vegetarian café on site. And don't skip your chance to visit the nunnery next door, which was founded in 1934, but also has Tang-style architecture (the campus was rebuilt to incorporate this architectural mode in 1998). It's one of the largest handmade wooden buildings in the world, and, impressively, it contains no nails in its construction. The nunnery's courtyards are filled with bonsai gardens, incense urns, and flowering bougainvillea.

www.nanliangarden.org +852 3658 9366

58 AMBER FORT

Devisinghpura, Amer, Jaipur, India

TO VISIT BEFORE YOU DIE BECAUSE

The opulent fort has a "floating" saffron garden.

One of six forts belonging to UNESCO's Hill Forts of Rajasthan World Heritage site, Amber Fort, sometimes called Amer Fort, is one of Jaipur's top attractions. The opulent Indo-Islamic palace was built into the hillside above the town of Amer in 1592, during the reign of Mughal emperor Akbar (1542–1605). Within Amber Fort itself are several beautiful courtyards with classic Mughal gardens—that is, in the *chahar bagh* style, separated into four sections by water features. But one of the best gardens is beneath the main palace, seemingly floating on Maota Lake below. This is the saffron garden, or Kesar Kyari, where three terraces on an artificial island are covered in geometric ornamental beds. It was reportedly built in the 15th century, meaning it predates the fort itself.

No official website

59 HUMAYUN'S TOMB

Nizamuddin East, New Delhi, India

TO VISIT
BEFORE YOU DIE
BECAUSE

As the first monumental garden tomb in India, Humayun's Tomb inspired the architect of the Taj Mahal.

Eight decades before the Taj Mahal was raised, Humayun's Tomb became the first large-scale garden tomb built in India—it would inspire the architect who would eventually design the subsequent mausoleum. When the tomb was constructed in the mid-16th century, its "grandeur of design and garden setting [had] no precedence in the Islamic world," according to UNESCO, which awarded it World Heritage site status in 1993. As was typical of Mughal landscape design at the time, the 67 acres of gardens surrounding the main building were designed in the classic *chahar bagh* style, with flowing water separating each plot into four sections, representing the four rivers of heaven as described in the Quran. The final resting place of Mughal emperor Humayun, which was funded by his son, Emperor Akbar, would go on to hold 150 members of the family, earning the site the moniker "dormitory of the Mughals."

No official website

60 MEHTAB BAGH

Near Taj Mahal, Agra, India

TO VISIT BEFORE YOU DIE BECAUSE

It provides the best off-site views of the Taj Mahal.

If you want the best view of the Taj Mahal, cross the Yamuna River and head for the Mehtab Bagh, or Moonlight Garden, which sits directly opposite the iconic mausoleum. The original garden was laid out by Emperor Babur (1483–1530), founder of the Mughal Empire, in the early 16th century—before the Taj Mahal was built. But after the tomb's construction, in the early to mid-17th century, Emperor Shah Jahan (1592–1666) renovated the *chahar bagh* gardens to be the ideal place from which to view the marble monument. In fact, the Mehtab Bagh's dimensions align perfectly with those of the Taj Mahal just across the river, providing a spectacular display of perspective. In the centuries following its renovation, the garden fell victim to frequent flooding, and for many years, it was just a pile of sand. But archaeologists discovered the site in the 1990s, and a long restoration process is returning it to its former grandeur.

No official website

61 ROCK GARDEN OF CHANDIGARH

Sector No.1, Chandigarh, India

TO VISIT BEFORE YOU DIE BECAUSE

One of the world's largest folk art sculpture parks, artist Nek Chand's garden was built in secret.

While Swiss-French architect Le Corbusier was developing a Modernist utopia for the city of Chandigarh in the 1960s, an unassuming, low-level government worker was building his own utopia nearby. During his day job as a transportation inspector, Nek Chand began collecting rocks, trash, and other debris from around town, including from Le Corbusier's construction site. Then, by cover of night, he entered a nearby forest—a protected one—and created little sculptures. Over 10 years, Chand's art garden grew massive in scale, until it was discovered by his government colleagues in 1976. But instead of demolishing the illegally built site, they provided the self-taught artist with a salary and a staff to continue his work. Today, the garden spans 25 acres and is perhaps even more beloved than Le Corbusier's construction.

www.nekchand.com/visit-rock-gardens +91 172 740 645

62 ERAM GARDEN

Eram Street, Shiraz, Iran

TO VISIT BEFORE YOU DIE BECAUSE

A designated UNESCO site, this garden is commonly regarded as one of the most beautiful places in Shiraz.

As with many historic sites around the world, Eram Garden's long story is a little muddled, but that doesn't detract from the splendid Qavam House and its associated garden that exist today. The original garden, which features the classic Persian *chahar bagh* quadripartite style (that is, divided by water into four sections representing the four Zoroastrian elements), is thought to have been laid out between the 11th and 13th centuries, while the major restoration that brought about the basis of the structures we see today occurred in the 18th century. Though the garden continued to undergo changes in its more recent history, Eram is still a premier example of a classic Persian garden—that's why it's one of nine gardens in Iran to receive a UNESCO World Heritage site designation under the shared entry "Persian Garden."

No official website

63 FIN GARDEN

Amir Kabir Street, Kashan, Iran

TO VISIT
BEFORE YOU DIE
BECAUSE

Built in 1590, this is the oldest existing garden in Iran.

UNESCO-designated sister garden to Eram, Fin was completed in 1590 for Safavid Shah Abbas the Great of Persia (1571–1623) and restored and expanded in the 19th century during the Qajar period to include extraordinary architecture, such as the recreational pavilion and the two-story pool house, and the teahouse. The pools that separate the garden's four sections, as well as its fountains, are fed by warm water that bubbles up from natural springs—that's why there's a popular hammam here. That bathhouse, however, is the site of an infamous murder. While soaking in a bath, Amir Kabir (1807–1852), formerly chief minister to Naser al-Din Shah Qajar (1849–1896), was killed by order of the shah. Beyond the gardens themselves, the complex is known for the small Kashani National Museum, which holds a collection of artifacts, from textiles to ceramics.

No official website

64 BAHÁ'Í GARDENS HAIFA

Yefe Nof Street 61, Haifa, Israel

TO VISIT BEFORE YOU DIE BECAUSE

With 19 Persian-style terraces culminating in the gold-domed Shrine of the Báb, this pilgrimage site is not one to miss.

Set atop Mount Carmel, the impressive Shrine of the Báb is one of the holiest sites of the Bahá'í Faith, a modern religion born in the 19th century. Leading up the hill are 19 highly decorated garden terraces that draw the attention not only of worshippers making the pilgrimage, but also of tourists to the Israeli city of Haifa. And that's intentional. Working off the initial designs of Bahá'í leader Shoghi Effendi (1897–1957), whose work remained only partially built for decades, Iranian architect Fariborz Sahba (b. 1948) created a layout that would visually pull visitors up to the shrine, using elements like water features, hedges, colorful flower beds, and architectural embellishments to complete the effect. The dazzling project, now a UNESCO World Heritage site, was completed in 2001 and is visited by more than a million people each year.

www.ganbahai.org.il +972 4 831 3131

65 ACROS FUKUOKA PREFECTURAL INTERNATIONAL HALL

1-1-1 Tenjin, Chuo-ku, Fukuoka, Japan

TO VISIT BEFORE YOU DIE BECAUSE

The roof garden of this Japanese office and cultural center is one of the first masterpieces of green design.

If residents of Fukuoka, Japan, were worried when a park in the city center was to be developed into a building, they needn't have been. The ACROS (Asian Cross Roads Over the Sea) Fukuoka Prefectural International Hall, a cultural hub built in 1994, is one of the best (and earliest) examples of green architecture in the world. Conceived by architect Emilio Ambasz (b. 1943), the building has a tiered roof known as the "Step Garden," where 15 terraces are filled with more than 50,000 plants; it seamlessly flows into Tenjin Central Park next door in a perfect symbiosis of architecture and green space. Beyond recreational use, the Step Garden is also an energy-saving technology for the building, as it helps regulate the temperature inside. And per a study by Kyushu University, it also helps combat the urban heat island effect in the neighborhood.

www.acros.or.jp/english

+81 92-725-9111

66 ADACHI MUSEUM OF ART

320 Furukawa-cho, Yasugi, Japan

TO VISIT BEFORE YOU DIE BECAUSE

Gardening might be considered an art form, but here, the gardens have actually become “living paintings.”

Zenko Adachi (1899–1990) had two loves in life: art and gardens. So, in 1970, he opened the Adachi Museum of Art in his rural hometown of Yasugi. There he displayed his collection of modern and contemporary Japanese paintings—including many works by Yokoyama Taikan (1868–1958), a leader of *Nihonga* painting—in a museum surrounded by exquisite traditional gardens, including a dry landscape garden and a moss garden. Unlike most Japanese gardens, however, they’re not meant to be entered, but to be seen from inside the museum, as if the windows were canvases hanging on the walls. The concept stems from Adachi’s belief that gardens should be appreciated as “living paintings.” As of 2021, the Adachi Museum of Art gardens have been named the best Japanese garden in the world for 18 consecutive years by *Sukiya Living Magazine*, also known as *The Journal of Japanese Gardening*.

www.adachi-museum.or.jp +81 0854-28-7111

67 HŌKOKU-JI

2-7-4 Jomyoji, Kamakura, Kanagawa, Japan

TO VISIT BEFORE YOU DIE BECAUSE

There's a reason this is known as the "Bamboo Temple."

Hōkoku-ji was built in 1334 as the family temple of the Ashikaga and Uesugi clans; it's a pretty modest complex with a main hall housing a statue of Buddha, a bell tower, and several auxiliary structures set behind a small garden with moss-covered stones and fragrant cherry and jasmine trees. But the real star of the show is in the back of the property, where a grove of more than 2,000 bamboo stalks creates a mysterious, serene atmosphere—that's why this site is nicknamed the "Bamboo Temple." Follow the winding paths through the forest, and you'll come upon a small teahouse. Elsewhere on the property are a series of small stone caves that are rumored to hold the ashes of Ashikaga leaders.

www.houkokuji.or.jp +81 46-722-0762

68 HOTEL CHINZANSO TOKYO

10-8, Sekiguchi 2-chome, Bunkyo-ku, Tokyo, Japan

TO VISIT BEFORE YOU DIE BECAUSE

The Hotel Chinzanso's historic garden is the definition of an urban oasis.

The name *Chinzanso* loosely means "villa on a mountain of camellias," which is a perfect descriptor of the Hotel Chinzanso Tokyo. Though the main building and current iteration of the garden date back to the Meiji era, or the late 19th century, the hilltop site has been known for its camellias for at least 700 years—Japanese artist Utagawa Hiroshige (1797–1858) even captured the site in his One Hundred Famous Views of Edo *ukiyo-e* woodblock prints. Those camellias still fill the 17-acre garden today, as do cherry blossoms, deciduous trees, and Japanese antiques like stone lanterns. There's also a three-story pagoda, an arched bridge, and an Inari shrine on the site.

www.hotel-chinzanso-tokyo.com/garden +81 3-3943-1111

69 KAIRAKU-EN

1-1251 Migawa, Mito, Ibaraki, Japan

TO VISIT BEFORE YOU DIE BECAUSE

More than 3,000 plum trees blossom here in late winter.

While most of Japan's top gardens were originally designed to be private spaces, Kairaku-en was built with the public in mind. The 23-acre garden, whose name loosely means "park that can be enjoyed together" or "park where everyone can relax," was constructed by order of *daimyō* Tokugawa Nariaki (1800–1860) of Mito in 1842, making this relatively young among Japan's major classical gardens. (Kairaku-en is one of the Three Great Gardens of Japan.) Nariaki intended it as a place of respite for students of his adjacent martial arts school and library but opened it to members of the samurai class on a regular basis, as well as the general public on a limited basis. It is best known for its 3,000-plus plum trees, which blossom in late February and early March, but it also has cherry blossoms, azaleas, and wisteria that bloom later in the year.

www.ibaraki-kairakuen.jp +81 029-244-5454

70 KENROKU-EN

1 Kenrokumachi, Kanazawa, Ishikawa, Japan

TO VISIT
BEFORE YOU DIE
BECAUSE

This is the ultimate Japanese landscape garden.

If you can visit only one garden in Japan, this should be it. Kenroku-en's name translates to "the garden of six sublimities," referring to the six principles of ancient Chinese landscape design philosophy that define a perfect garden: spaciousness and seclusion, artifice and antiquity, and waterways and panoramas. Naturally, this garden has all six. Built over hundreds of years by the Maeda clan as their private landscape garden on the outer grounds of Kanazawa Castle, Kenroku-en is a strolling-style garden, designed to be walked through rather than sat in. As you explore its curving paths that wind through its 28-acre grounds, you'll pass highlights like the Kotojitoro stone lantern, avenues of cherry blossoms, the Yūgao-tei teahouse (the oldest building in the garden, built in 1774), and the Karasaki pine tree. Kenroku-en has the distinction of being one of the Three Great Gardens of Japan, the others being Kairaku-en in Mito and Kōraku-en in Okayama.

www.pref.ishikawa.jp/siro-niwa/kenrokuen +81 76-234-3800

71 KŌRAKU-EN

Korakuen 1-5, Kita-ku, Okayama, Japan

TO VISIT BEFORE YOU DIE BECAUSE

Its sprawling lawns make it unique among Japanese gardens.

Set on an island in the Asahi River in Okayama, Kōraku-en joins Kairaku-en and Kenroku-en as one of the Three Great Gardens of Japan. It was built in 1700 for *daimyō* Ikeda Tsunamasa (1638–1714), who used it not only as a place for entertaining guests, but also as a private retreat for other feudal lords. Eventually, members of the public were also invited to visit the garden on limited occasions. Kōraku-en has the distinction of being very well recorded in both drawings and writings, allowing historians to trace the garden's development over the past three centuries; after damage during the 1934 floods and during World War II, restorers were able to rebuild the garden with heightened accuracy. Unlike many classical Japanese gardens, Kōraku-en has large lawns between its ponds, teahouses, and blossoming trees. Iconic American architect Frank Lloyd Wright (1867–1959), who was largely inspired by Japanese aesthetics, visited the garden in 1905 on his first trip to the country.

www.okayama-korakuen.jp +81 86-272-1148

72 MEIGETSU-IN TEMPLE

189 Yamanouchi, Kamakura, Japan

TO VISIT BEFORE YOU DIE BECAUSE

The hydrangea blossoms in June are a spectacle to behold.

Meigetsu-in was originally a sub-temple of the 12th-century Zenkō-ji in Kamakura, but after the Meiji Restoration in the 19th century closed the larger complex, it's all that remains. Sometimes referred to as the "Hydrangea Temple," Meigetsu-in is famous for its thousands of hydrangeas that bloom each June—the gardens become quite busy during this time and during a small window in the fall, in which visitors can enjoy the autumnal colors. They celebrate the life of their builder, Uesugi Norikata (1335–1394), founder of the Uesugi clan, who is thought to be buried in a cave in the garden, as is Hōjō Tokiyori (1227–1263), one of the most powerful leaders of the Kamakura shogunate.

www.trip-kamakura.com/place/230.html +81 467-24-3437

73 RYŌAN-JI

13 Ryoanji Goryonoshitacho, Ukyo Ward, Kyoto, Japan

TO VISIT BEFORE YOU DIE BECAUSE

This is Japan's preeminent rock garden.

The former capital city of Japan, Kyoto has such a rich cultural heritage that UNESCO gave 17 of its sites World Heritage status under the umbrella of "Historic Monuments of Ancient Kyoto." That includes Ryōan-ji, a Zen temple dating back to 1450 with one of the best examples of *kare-sansui*, or dry landscape, where rough boulders are surrounded by a field of smooth river rocks that are raked into smooth lines that represent flowing water. Ryōan-ji's rock garden has 15 boulders positioned so that you can never see them all, no matter where you stand in the garden—it's a visual riddle of sorts, asking the viewer to contemplate the layout of the garden. The composer John Cage (1912–1992) was so moved by the rock garden that he wrote a piece of music and created a series of drawings inspired by it.

www.ryoanji.jp/top.html

+81 75-463-2216

74 SAIHŌ-JI

56, Matsuo Jingatanichō, Nishikyō-ku, Kyoto, Japan

TO VISIT BEFORE YOU DIE BECAUSE

More than 120 types of moss grow in this Zen temple's verdant garden.

Founded in the middle of the 8th century, Saihō-ji is a Zen temple complex known for its mossy garden; there are more than 120 types of moss growing here, which is why it's referred to as *Koke-Dera*, or "Moss Temple." It's important to note that visits to the Saihō-ji and its garden are by reservation only to help manage crowds and protect the delicate mosses. Up until 2021, guests had to request appointments via mailed letter, but an online reservation system is under development. Upon arrival, visitors must participate in a ritual with the monks before entering the gardens, such as prayer, meditation, or writing sutras. They'll then be free to explore the grounds, which include the Ougonchi Pond and the Shōnantei teahouse. Like Ryōan-ji, this is one of the 17 UNESCO-designated Historic Monuments of Ancient Kyoto.

www.saihoji-kokedera.com +81 75-391-3631

75 TENRYŪ-JI

68 Susukinobaba-cho, Kyoto, Japan

TO VISIT BEFORE YOU DIE BECAUSE

This garden sits on the site of the first Zen temple in Japan.

Though the "modern" temple of Tenryū-ji was built by shogun Ashikaga Takauji (1305–1358) in 1339, with gardens designed by the famous monk Musō Soseki (1275–1351), its site has a far older relationship with Zen Buddhism. It was home to the first Zen temple in Japan, Danrin-ji, built by the Empress Tachibana no Kachiko (786–850) in the 9th century. While Tenryū-ji has been plagued by fires and wars that have destroyed many of its buildings over the last 700 years, a restoration campaign in the 20th century has brought back the site's greatness. The garden is particularly well regarded for its use of *shakkei*, or borrowed scenery; Soseki factored the hills beyond the garden's boundaries into his design. Of the 17 UNESCO-designated Historic Monuments of Ancient Kyoto, this is one of the most treasured Zen temples.

www.tenryuji.com +81 75-881-1235

76 NATIONAL KANDAWGYI BOTANICAL GARDENS

Nandar Street, Ward 6, Pyin Oo Lwin, Mandalay, Myanmar

TO VISIT BEFORE YOU DIE BECAUSE

This is the premier destination in Myanmar to learn about the country's flora—and to see a dazzling flower show.

In 1915, the government of Myanmar set aside 30 acres of land to display the country's native plants; today, that garden is the National Kandawgyi Botanical Gardens, and it covers 437 acres. The gardens' major transformation began in 1917, under the guidance of forest researcher Alex Rogers (1864–1937) and amateur botanist Lady Charlotte Wheeler-Cuffe (1867–1967), who set out to design gardens reminiscent of the Kew Gardens in London. Today the National Kandawgyi Botanical Gardens are the main ecotourism site in Myanmar, and they still have a focus on native flora—one of the highlights of the collection is the garden of indigenous orchids. But the biggest event at the gardens is the annual flower festival, which showcases the work of the country's flower producers.

www.nationalkandawgyigardens.com +959 5048210

77 GARDEN OF DREAMS

Kaiser Mahal, Tridevi Marg, Kathmandu, Nepal

TO VISIT BEFORE YOU DIE BECAUSE

There aren't very many European-style public gardens in Nepal. In fact, this might be the only one.

You might not expect to find a neoclassical garden in the Nepalese capital of Kathmandu, but tucked away on a small one-acre plot is the Edwardian-style private garden of Nepalese field marshal Kaiser Shumsher Rana (1892–1964). Rana, having visited England on a number of occasions, not to mention having an extensive library, had landscape architect Kishore Narsingh replicate the Edwardian garden's blend of formality and naturalism, with a classic axial plan combined with naturalistic planting. When it was built in 1920, the garden had six pavilions, one dedicated to each of Nepal's seasons, though only three remain today. A restoration project by Nepal's Ministry of Education and the government of Austria ushered the garden into its second century in style.

www.gardenofdreams.org.np + 977 1-4425340

78 SHALIMAR BAGH

Shalamar Chowk, Shalamar Town, GT Road, Lahore, Pakistan

TO VISIT BEFORE YOU DIE BECAUSE

This is the pinnacle of Mughal garden design.

Completed in 1642 for Emperor Shah Jahan (1592–1666), about a decade after the Taj Mahal, Shalimar Bagh is one of the most magnificent Mughal gardens, if not the most magnificent, in the world. UNESCO, which awarded it World Heritage site status in conjunction with nearby Lahore Fort in 1981, suggests that the gardens bear "witness to the apogee of Mughal artistic expression." That expression translates to two terraces of four-part *chahar bagh* gardens—one for the emperor's harem and one for the general public—as well as a third terrace between them for his private use. Its standout feature is its complex water system, which includes hundreds of fountains, cascades, and canals.

www.shalimar-bagh.business.site +92 0312 4856085

79 CATHERINE PARK

Sadovaya Street, 7, Pushkin, St. Petersburg, Russia

TO VISIT BEFORE YOU DIE BECAUSE

The backdrop to this Dutch-style garden and English-style landscape park is the grandiose Catherine Palace.

While the Rococo Catherine Palace is a must-see for all visitors to St. Petersburg, don't miss the extraordinary Catherine Park adjacent to it. It features two sections: the Dutch-style Old Garden, and the English-style Landscape Park. The former was designed in the 1720s by Dutch master gardeners Jan Roosen and Johann Vocht, who developed the parterres and several ponds, and then expanded greatly in the decades following with sculptures by Bartolomeo Francesco Rastrelli (1700–1771), the architect of the palace itself, and structures by architect Vasily Neyolov (1722–1782). Neyolov also oversaw the development of the English-style landscape park in the 1770s, though it was completed by English gardener John Bush. For Catherine the Great, the reigning empress, the gardens were a way to demonstrate her attention to stylistic trends as well as her monarchical accomplishments. Considering that Catherine Park is still so highly regarded today, she succeeded.

www.tzar.ru/en/objects/ekaterininskypark +7 812-415-76-67

80 MAIN BOTANICAL GARDEN OF THE RUSSIAN ACADEMY OF SCIENCES

Botanicheskaya Street, 4, Moscow, Russia

TO VISIT BEFORE YOU DIE BECAUSE

Russia's preeminent botanical garden sprawls across 820 acres, making it one of the largest botanical gardens in Europe.

Set on the former hunting grounds of Tsar Alexei Mikhailovich (1645–1676), Main Botanical Garden of the Russian Academy of Sciences was established in 1945. Its first director was Russian botanist Nikolai Vasilevich Tsitsin (1898–1980), a leading researcher of plant genetics—to this day, the gardens are a leading research institute for experimental botany. But the grounds, which are home to 18,000 species of plants, are also open to visitors, who can explore a Japanese garden designed by landscape architect Takeshi "Ken" Nakajima (1914–2000), a rose garden with 20,000 rose bushes, an oak forest with century-old trees, and more than 100,000 square feet of greenhouses filled with 25,000 tropical and subtropical plants.

www.gbsad.ru +7 8-499-977-91-45

81 PETERHOF

Razvodnaya Ulitsa, 2, St. Petersburg, Russia

TO VISIT BEFORE YOU DIE BECAUSE

The fountains alone at this palace complex inspired by Versailles are enough reason to visit.

After a visit to Louis XIV's Versailles in France, Peter the Great (1672–1725) decided he wanted his own version of it at home in Russia—and that included the gardens. He hired Swiss architect Domenico Trezzini (c. 1670–1735) and French landscape designer Jean-Baptiste Alexandre Le Blond (1679–1719), who worked with Versailles' landscape master André Le Nôtre to build it—and the rest of St. Petersburg—for him. Peterhof's gardens are divided into several sections, including the Lower Park, the Upper Park, and the breathtaking Grand Cascade, a monumental water complex of 64 fountains and 225 sculptures. (There are more than twice that number of fountains throughout the entire estate.) Impressively, there are no pumps to power those water features; water is sourced from natural springs and flows through a gravity-based system to power the displays.

www.peterhofmuseum.ru +7 812-313-23-14

82 GARDENS BY THE BAY

18 Marina Gardens Drive, Singapore 018953, Singapore

TO VISIT BEFORE YOU DIE BECAUSE

Who wouldn't want to see the iconic Supertrees and cooled conservatories, which are among the most recognisable features of Singapore's skyline?

Comprising three distinctive waterfront gardens in the heart of Singapore's downtown, Gardens by the Bay is more than a nature park with over 1.5 million plants. It is also a recreational green space for the community, an "edutainment" complex, and home to a collection of interesting art sculptures! Bay South Garden, opened in 2012, is the largest of the three gardens and likely the one you're most familiar with. It's home to the iconic Supertrees as well as the Cloud Forest and Flower Dome conservatories, which are cooled rather than heated to provide a salubrious climate for a diverse collection of plants from around the world, enabling them to thrive in Singapore's hot, humid climate. The Flower Dome is also the world's largest glass greenhouse, with a volume equivalent to 75 Olympic-sized swimming pools! The other two gardens that make up Gardens by the Bay are Bay East and Bay Central. Bay East offers a tranquil respite from the bustling city and stunning views of the Singapore skyline, while Bay Central, when it is developed, will boast a 3km waterfront promenade, from which visitors can enjoy scenic views of the city.

www.gardensbythebay.com.sg +65 6420 6848

83 JEWEL CHANGI AIRPORT

78 Airport Boulevard, Singapore

TO VISIT BEFORE YOU DIE BECAUSE

It's hard to believe this otherworldly garden complex—complete with a sky waterfall—is inside an airport.

Most people see airports as throughpoints on their journeys. But Jewel Changi Airport is here to change your perspective on transportation hubs. More of a nature-themed entertainment and shopping center than a collection of gates and food courts, Jewel is a wonderland of indoor gardens surrounding the HSBC Rain Vortex, the world's tallest indoor waterfall. The primary garden is the Shiseido Forest Valley, a multisensory oasis with walking trails through lush greenery, but there's also a hedge maze, a topiary garden, and a colorful canopy garden for families. Best of all, everything is indoors, meaning you're protected from Singapore's hot and humid climate—it feels like something out of science fiction, like a greenhouse on Mars. This area of the airport is located on the landside, not the airside, meaning you don't need a plane ticket to enjoy the attractions.

www.jewelchangiairport.com/en/attractions/forest-valley.html

+65 6956 9898

84 SINGAPORE BOTANIC GARDENS

1 Cluny Road, Singapore

TO VISIT BEFORE YOU DIE BECAUSE

This is Singapore's first UNESCO World Heritage site and the only tropical botanical garden with the designation.

The very first national garden in Singapore was planted in 1822 by Sir Stamford Raffles (1781–1826), the founder of modern Singapore, who wanted to test crops for cultivation. In 1859, a new national garden was developed: the Singapore Botanic Gardens. In the first several decades of its existence, the gardens continued Raffles' mission, successfully cultivating the Pará rubber tree from seeds sent by London's Kew Gardens, which became one of Singapore's most bountiful crops. But it was also a leisure garden with landscape design that matched that of the English Landscape Movement. Today the gardens span more than 200 acres and are lauded for the 7.5-acre National Orchid Garden. The Singapore Botanic Gardens were the first site in Singapore to be awarded UNESCO World Heritage site status and the only tropical botanical garden currently on the list.

www.nparks.gov.sg/sbg +65 1800-471-7300

85 BIWON SECRET GARDEN

99, Yulgok-ro, Jongno-gu, Seoul, South Korea

TO VISIT BEFORE YOU DIE BECAUSE

These were once the private gardens of Changdeokgung Palace.

Also known as the Huwon Garden, the Biwon Secret Garden is set behind 15th-century Changdeokgung Palace, a UNESCO World Heritage site. (Huwon means "rear" and Biwon means "secret.") It's hardly a secret, though, since you can book guided tours of the grounds—and you should do so in advance of your visit, as they're required. But its name likely derives from the fact that the 78-acre garden was once leisure space for the Joseon Dynasty's royal families who lived in the palace. Unlike other formal gardens around the world, Biwon was kept as natural as possible, with gardeners advised not to touch anything unless absolutely necessary. But it's not entirely wild, as there are historic pavilions, halls, and gates throughout the garden.

www.cdg.go.kr +82 2-3668-2300

86 GARDEN OF MORNING CALM

432, Sumogwon-ro, Gapyeong-gun, Gyeonggi-do, South Korea

TO VISIT
BEFORE YOU DIE
BECAUSE

The garden hosts festivals all year long, from exhibits dedicated to flowers to winter displays of lights.

After visiting gardens around the world, horticulture professor Sang-Kyung Han of Sahmyook University decided to build a garden outside of Seoul that would represent Korean ideals of beauty. He established the Garden of Morning Calm, referring to the name given to Korea by Indian poet Rabindranath Tagore (1861–1941), “Land of the Morning Calm.” In the early days, the garden was divided into 10 thematic sections; today, it has 22 sections on 8.5 acres, with a living collection of 5,000 species of plants. The garden is known for its many festivals, which are held throughout the year, perhaps most famously the Lighting Festival in the winter, in which thousands of colorful lights illuminate the plants.

www.morningcalm.co.kr +82 1544-6703

87 MAGUL UYANA

Anuradhapura, Sri Lanka

TO VISIT BEFORE YOU DIE BECAUSE

This garden is 2,500 years old.

In a way, the Sacred City of Anuradhapura has its roots in gardening—it was built around a cutting from the Buddha's Bodhi tree in the 3rd century BCE. The ancient capital city of the Sinhalese Anuradhapura Kingdom was one of the most important metropolises in the region for some 1,300 years, but it was invaded in 993, then abandoned. Though it remained deserted for centuries, it has since been repopulated as a small town around ruins of the ancient city, including the Magul Uyana, or royal gardens. Set in the present-day Ranmasu Uyana park, the remnants of the royal gardens include ancient bathing pools, which had an impressive hydraulics system for pumping water, as well as a strange chart-like carving known as the Sakwala Chakraya. Some theorize the carving indicates a "stargate," or a portal created by extraterrestrials, but archeologists say there's no evidence to support this. Regardless of whether or not there has been alien contact here, UNESCO has designated Anuradhapura, including the Magul Uyana, a World Heritage site.

No official website

88 ROYAL BOTANIC GARDENS, PERADENIYA

Peradeniya Road, Kandy, Sri Lanka

TO VISIT BEFORE YOU DIE BECAUSE

These are the most elaborate botanic gardens in Sri Lanka, with a history that dates back 700 years.

Though the Royal Botanical Gardens of Perandeniya were formally founded by Alexander Moon (?–1825) in 1821, there have been gardens on this site since 1371, when King Wickramabahu III established his royal gardens there, which remained in use under different rulers through the 18th century. Over the past century, as a formal botanic garden, the grounds have expanded to cover nearly 150 acres, and they're home to both native and nonnative flora, including the cannonball tree planted by King George V (1865–1926) and Queen Mary (1893–1936) at the turn of the 20th century. Highlights of the collection include the Orchid House, where visitors can see *Grammatophyllum speciosum*, the largest species of orchid in the world, as well as a massive—and oft-photographed—Javan fig tree.

www.botanicgardens.gov.lk +94 812 388 088

89 SIGIRIYA

8 Ela pahalawewa, Dambulla 21100, Sri Lanka

TO VISIT BEFORE YOU DIE BECAUSE

The gardens at the “Lion’s Rock” fortress were built in the 5th century.

The aggressive King Kashyapa I, who ruled the native Sinhalese Moriya Dynasty, built his imposing clifftop-fortified city Sigiriya in the late 5th century, after he overthrew and executed his father to take the throne. Despite this brutal history, the gardens of Sigiriya, both at the base of the rock (an ancient volcanic plug that juts out from the jungle) and atop it, are some of the most beautiful and well-preserved ancient gardens in the world. They primarily follow the rectilinear *chahar bagh* style, and their flowing water features are impressively still operational today. Sigiriya, which means “Lion’s Rock,” so named for the lion gateway carved into the rock face at the entrance to the city, was abandoned after the king’s death, though it later served as a monastery. UNESCO designated it a World Heritage site in 1982.

No official website

90 MAE FAH LUANG GARDEN

Doi Tung Development Project, Mae Fa Luang, Mae Fah Luang District, Chiang Rai 57240, Thailand

TO VISIT BEFORE YOU DIE BECAUSE

The bright flower beds in the middle of the forest look like splashes of paint on a green canvas.

The well-educated, well-traveled, and well-loved Princess Srinagarindra (1900–1995), also known as the Princess Mother, dedicated her own life to improving the lives of others. She worked tirelessly to provide services to the minority groups living in the remote mountains in northern Thailand, flying in supplies and even medical teams on helicopters. Because of this, the locals gave her the name Mae Fah Luang, or "Royal Mother from the Sky." She founded the Mae Fah Luang Foundation (MFLF) to expand her efforts; the nonprofit organization provides education and training through the Sustainable Alternative Livelihood Development program. One of the MFLF's projects is the Mae Fah Luang Garden, developed per the Princess Mother's wishes to create a flower garden for Thais—she was known for her love of flowers. The colorful garden, which is filled with rare orchids, has provided sustainable livelihood for many locals, who are trained in horticulture so they can maintain the grounds.

No official website +66 2 252-7114

91 NONG NOOCH TROPICAL GARDEN

34 Na Chom Thian, Sattahip District, Pattaya, Thailand

TO VISIT BEFORE YOU DIE BECAUSE

If there were such a thing as a garden theme park, this is it.

Covering more than 650 acres, Nong Nooch Tropical Garden can overwhelm with its size, much less its exuberant floral displays and funky installations. After buying acres of rolling hills outside Pattaya, Thailand, in 1954, Pisit and Nongnooch Tansacha intended to transform their newly purchased land into a fruit plantation. But after a trip abroad, Mrs. Tanascha was inspired by the gardens she saw to create her own ornamental masterpiece—one that doubled as a conservation center. Named in her honor, the Nong Nooch Tropical Garden opened to tourists in 1980 with themed gardens, including a dinosaur garden, a French garden, a Stonehenge-inspired garden, a bonsai garden, and an orchid garden. The park also has cultural shows and attractions, as well as an on-site resort with accommodations, restaurants, and a spa. But this is also a research facility, and it's home to one of the world's largest cycad collections, as well as a cycad gene bank.

www.nongnoochtropicalgarden.com +66 81 919 2153

92 DUBAI MIRACLE GARDEN

Al Barsha South 3, Dubailand, Dubai, United Arab Emirates

TO VISIT BEFORE YOU DIE BECAUSE

There aren't many places on earth where you can see 150 million flowers at once. This garden is one of them.

Everything is bigger and flashier in Dubai, and that includes its gardens. The Dubai Miracle Garden is indeed a miracle as its name suggests—despite its location in the desert, the 18-acre space is home to more than 150 million flowers. While water is a precious commodity here, the garden uses cleaned wastewater to provide hydration for its overwhelming number of blooms. And because typical flower beds won't do in Dubai, the flowers are arranged into massive installations, from a neighborhood of floral villas to a giant teddy bear. The garden also houses the world's largest floral installation: a life-size replica of the double-decker Airbus A380 aircraft, complete with the livery of Dubai's own carrier, Emirates. You can also find the world's tallest supported topiary structure, a 59.25-foot-tall Mickey Mouse, on the grounds.

www.dubaimiraclegarden.com

no tel. no. available

93 BELVEDERE GARDENS

Prinz Eugen-Straße 27, Vienna, Austria

TO VISIT BEFORE YOU DIE BECAUSE

The exquisite Baroque gardens are the glue that joins the two Belvedere palaces.

The Belvedere Museum complex in Vienna, a UNESCO World Heritage site, has something of an unusual layout: two 18th-century palaces sit opposite one another on a gentle slope. Between them, however, is one of the most celebrated Baroque gardens in Europe. Spread across three terraces with flowing fountains, the Belvedere's formal French palace gardens were designed by Dominique Girard (c. 1680–1738), a student of the Château de Versailles' landscape architect André Le Nôtre (1613–1700). The Belvedere is also home to several other gardens, including the more intimate Kammergarten (Privvy Garden) and the oldest alpine garden in Europe. But the Belvedere is also a contemporary institution: at the nearby Belvedere 21 venue is a sculpture garden with rotating exhibitions.

www.belvedere.at/en/museum/gardens +43 1 795 57-0

94 SCHLOSSPARK SCHÖNBRUNN

Schönbrunner Schloßstraße 47, Vienna, Austria

TO VISIT BEFORE YOU DIE BECAUSE

This is the Habsburg Dynasty's response to Versailles—and it's truly magnificent.

Like a phoenix, the UNESCO-designated gardens at Schönbrunn have completed several cycles of life, death, and rebirth. The original gardens, part of what was known as the Katterburg estate of Emperor Maximilian II (1527–1576), were destroyed by the Hungarians in 1605. They were rebuilt under Eleonora of Gonzaga (1598–1655) and her niece, also Eleonora of Gonzaga (1630–1686), then destroyed by the Ottomans in 1686. In 1694, landscape architect Jean Tréhet (1654–1740), a student of Versailles' master garden designer André Le Nôtre, designed the Baroque gardens that have survived to the present day under Joseph I, with many restorations and expansions by Maria Theresa (1717–1780). Schlosspark Schönbrunn is now home to a French formal garden, an orangery, a gloriette, a privy garden, a palm house, a desert house, multiple fountains, and, of course, the Tiergarten Schönbrunn, the former royal menagerie that is the world's oldest zoo.

www.schoenbrunn.at +43 1 811 13-0

95 MEISE BOTANIC GARDEN (PLANTENTUIN MEISE)

Nieuwelaan 38, Meise, Belgium

TO VISIT BEFORE YOU DIE BECAUSE

Covering 227 acres, the Meise Botanic Garden is one of the largest botanical gardens in Europe.

Formerly known as the National Botanic Garden of Belgium, the Meise Botanic Garden was formally established in Brussels in 1840, though its roots go back to 1796. But it moved to its current, more spacious home outside the city in 1958, on the grounds of the historic Bouchout Castle. Across its 227 acres (including 2.5 acres of greenhouses) are 18,000 species of plants that represent about 6 percent of all known flora on Earth, including rare and endangered plants like the Kwango giant cycad. Another highlight, particularly for caffeine lovers, is Meise Botanic Garden's coffee plant collection. The garden also has an enormous botanical library with more than 200,000 books, the oldest of which were printed in the 15th century.

www.plantentuinmeise.be/en/home/ +32 02 260 09 70

96 PARC DES TOPIAIRES

Haie Himbe Street, 1, Durbuy, Belgium

TO VISIT BEFORE YOU DIE BECAUSE

This garden brands itself as the world's largest topiary garden in the world's smallest city.

While the town of Durbuy, Belgium, calls itself the "smallest city in the world," that might not actually be true—but given its petite size and population, it does have a rather impressively large topiary garden set along the riverbank. The Parc des Topiaires is about 2.5 acres in size, and it's home to some 250 topiary sculptures, ranging from geometric forms, to animals, to Dolly Parton reclining on a chair (or at least that's who one reviewer claims the topiary's model is). The topiaries are all carved out of boxwood shrubs, the oldest of which took root some 120 years ago. Open to the public since 1997, the garden also has a collection of medicinal plants on view, and it hosts bonsai workshops once a month.

www.topiaires.be +32 86 219 075

97 PALACE GARDENS BELOW PRAGUE CASTLE

Valdštejnská 158/14, Prague, Czech Republic

TO VISIT BEFORE YOU DIE BECAUSE

This patchwork of hillside gardens is a living example of Baroque landscape design.

Tucked into the slope between the great Prague Castle and Valdštejnská Street are five terraced gardens: Ledebour Garden, Small and Great Pálffy Gardens, Kolowrat Garden, and Small Fürstenberg Garden, each named for the former owners of the townhouses they sit behind. These comprise the Palace Gardens Below Prague Castle, compact yet ornate examples of Baroque garden design, where architectural decorations (think elegant staircases and ornamented pavilions), artwork (think murals and sculptures), and, of course, plenty of plants combine to create romantic spaces. Though archeological research has determined the existence of gardens on these grounds as early as the 13th century, the gardens in their current iterations were largely built between the 17th and 18th centuries.

www.palacove-zahrady.cz/en +420 257 148 817

98 PETŘÍN GARDENS

Prague, Czech Republic

TO VISIT BEFORE YOU DIE BECAUSE

From these series of parks and gardens, you can take in the best views of Prague's city center and Prague Castle.

Start your journey to this hillside recreation area by hopping on the funicular from Malá Strana; it's much easier to walk downhill than uphill! At the top is the iconic Petřín Lookout Tower, modeled after the Eiffel Tower in Paris, from which you can enjoy panoramic views of Prague. It's worth a visit as you make your way to the gardens elsewhere on the hill. As you descend, stroll through the Rosarium, where more than 3,000 roses bloom each summer; the Seminar Garden, a peaceful garden of fruit trees that was formerly part of a Carmelite monastery; and the Nebozízek Garden, a romantic orchard that surrounds the statue of poet Karel Hynek Mácha (1810–1836). At the base of the hill is the English-style Kinský Garden, and though it's technically separated from the Petřín Gardens by the Hunger Wall, a former fortification, they're also worth a stop while you're in the neighborhood.

www.prague.eu/en/object/places/504/petrin-gardens-petrinske-sady

no tel. no. available

99 FREDERIKSBORG CASTLE

Frederiksborg Slot 10, Hillerød, Denmark

TO VISIT BEFORE YOU DIE BECAUSE

The Renaissance castle and its Baroque gardens are set on a lake and they look as if they belong in a fairy tale.

Though Frederiksborg Castle itself was built in the 17th century and features grand Renaissance architecture, its gardens, designed by architect J. C. Krieger (1683–1755), came a little later, in 1720. The main garden took on the sumptuous Baroque style, with a focus on symmetry, as seen through the parterres and meticulously trimmed hedges, while the English garden has a more relaxed, romantic atmosphere. After the 18th century, the gardens fell into disrepair, until they were expertly restored to their original appearance in the 1990s. Today, the castle is home to the Museum of National History, which opened there in 1878, and visitors are encouraged to stroll the gardens.

www.dnm.dk/en/frederiksborg-castle +45 48 26 04 39

100 UNIVERSITY OF COPENHAGEN BOTANICAL GARDEN

Gothersgade 128, Copenhagen, Denmark

TO VISIT BEFORE YOU DIE BECAUSE

The garden has an impressive complex of 27 greenhouses anchored by the 1874 Palm House.

The University of Copenhagen Botanical Garden, part of the Natural History Museum of Denmark, has a history dating back to the year 1600, but its current iteration was established in 1870 after moving locations several times. Four years later, the elaborate Palm House was constructed, followed by 26 other glasshouses that now house the majority of tropical and subtropical plants, including the *Amorphophallus titanum* corpse flower, giant bamboo, and a significant collection of orchids. Architect Peter Christian Bønnecke (1841–1914), at the urging of brewer J. C. Jacobsen (1811–1887) of Carlsberg fame, modeled the building after the famous Crystal Palace built in London for the 1851 Great Exhibition. Today the garden holds more than 10,000 species of plants in both its conservatories and in outdoor flower beds.

www.snm.ku.dk/english/botanical-garden +45 22 83 65 97

101 SAPOKKA WATER GARDEN

Tallinnankatu 12, 48130 Kotka, Finland

TO VISIT BEFORE YOU DIE BECAUSE

At this serene harborside park, a 66-foot-tall waterfall is surrounded by flowers.

Hamina-Kotka might be one of the busiest cargo ports in Finland, but there's an oasis of tranquility to be found right off the bustling harbor. The Sapokka Water Garden, which underwent a major renovation in 1994, is centered on a 66-foot-tall waterfall, which ultimately empties out into the bay (and, in fact, oxygenates it, maintaining a friendly environment for the carp that keep the water clean). Planted throughout the garden are hundreds, if not thousands, of flowering plants—visit in the spring to see the onion flowers in bloom and in the summer to see the roses and azaleas in all their glory. In the fall, autumnal colors sweep through the park, while in the winter, the frozen waterfall takes on a sculptural quality, particularly when illuminated with spotlights. Throughout the park are sculptures depicting animals native to the area.

www.kotka.fi/asuminen-ja-ymparisto/puistot-ja-viheralueet/puistot/sapokan-vesipuisto

no tel. no. available

102 CHÂTEAU DE VAUX-LE-VICOMTE

Maincy, France

TO VISIT BEFORE YOU DIE BECAUSE

The project was the precursor to the Château de Versailles.

While the career highlight of architect Louis Le Vau, landscape architect André Le Nôtre, and artist Charles Le Brun is undoubtedly the Château de Versailles, their partnership might not have come about if it weren't for the Château de Vaux-le-Vicomte. Young politician Nicolas Fouquet (1615–1680) hired the trio to build his estate in Maincy, outside of Paris, in 1641. It was their first large project together, and it marked the debut of the opulent Louis XIV style. As for the gardens, Le Nôtre achieved a masterpiece considered to be the first true example of the *jardin à la française*, or French formal garden, defined by symmetry, order, and harmony between nature and the built environment. At Vaux-le-Vicomte, a nearly 2-mile-long axis anchors the garden to the château, and branching off from it are artistic parterres, water features, and trimmed shrubbery.

www.vaux-le-vicomte.com

no tel. no. available

103 CHÂTEAU DU CHAMP DE BATAILLE

8 Route du Château, Sainte-Opportune-du-Bosc, France

TO VISIT BEFORE YOU DIE BECAUSE

Honored as a *Jardin remarquable* by the French Ministry of Culture, this grand private garden being restored by designer Jacques Garcia is an under-the-radar treasure.

When renowned interior designer Jacques Garcia (b. 1947) bought the 17th-century Château du Champ de Bataille in Normandy in 1992, he wasn't only interested in renovating the interiors. Behind the stately home was a massive garden in disrepair, which Garcia has spent the last few decades delicately restoring to its original grandeur, and continues to do today. The 100-plus-acre garden is quintessentially French Baroque with immaculately manicured topiaries, ornate fountains, and sculptures and architecture dedicated to the Antique—it's so grand that it feels like it could be a little sister to Versailles. But the best part is that, while the Champ de Bataille garden is open to the public, it never has excessive crowds.

www.chateauduchampdebataille.com +33 2 32 34 84 34

104 CHÂTEAU ET LES JARDINS DE VILLANDRY

3 Rue Principale, Villandry, France

TO VISIT
BEFORE YOU DIE
BECAUSE

The gardens of this Renaissance château are among the most highly regarded in France.

Most châteaux belonged to royalty or the aristocracy, but the Château de Villandry is not most châteaux. It was the residence of Jean Le Breton (?–1556), the Minister of Finance for François I, who razed the existing fortress on the property when he purchased it to build the Renaissance palace that exists today. Le Breton wasn't a stranger to the construction of châteaux—he supervised the building of the resplendent Château de Chambord for the king. He also developed a love for gardening during his time as the ambassador to Rome, which he put to use at Villandry. Two centuries later, the gardens were restored and expanded upon by the Marquis of Castellane (1703–1782), who incorporated into them the French formal style of the time. The final restoration began in 1906, when Joachim Carvallo (1869–1936) bought Villandry and set out to replicate its Renaissance splendor.

www.chateauvillandry.fr + 33 2 47 50 02 09

105 FONDATION CLAUDE MONET

84 Rue Claude Monet, Giverny, France

TO VISIT BEFORE YOU DIE BECAUSE

This is the real-life setting of many of Claude Monet's most iconic Impressionist paintings.

French painter Claude Monet (1840–1926) wasn't just an artist with a brush; he was an artist with a trowel, too. The Impressionist master was always drawn to the outdoors in his art, often opting to work *en plein air*, or outdoors, rather than in a studio, and he was known to plant his own gardens that would serve as his paintings' subjects. His most famous is the water garden, home to hundreds of water lilies and the iconic green Japanese footbridge, at his home and studio in Giverny. He moved to the rural village in 1883, purchased the land for the water garden a decade later, and spent the rest of his career developing the gardens and painting his creation. Visitors can now stroll the gardens that captivated Monet, plus tour his home and studio for a glimpse into the life of the great painter.

www.fondation-monet.com +33 2 32 51 28 21

106 HANGING GARDENS OF MARQUEYSSAC (JARDINS SUSPENDUS DE MARQUEYSSAC)

Vézac, France

TO VISIT BEFORE YOU DIE BECAUSE

The sinuous topiaries have earned this garden a spot on France's *Jardins remarquables* list.

While the 17th-century Château de Marqueyssac in southwestern France originally had a French formal garden, gardener Julien de Cerval (1818–1893) decided to dramatically change the look of the grounds when he purchased the property in 1861. Instead of following the symmetry and order of the classic French formal style, de Cerval planted tens of thousands of boxwood shrubs, sculpting them into abstract rounded shapes that have a softer, more romantic aesthetic—a look that became much more popular under the rule of Napoleon III (emperor 1852–1870). Today, there are more than three and a half miles of winding paths meandering through an astonishing 150,000 sculpted boxwoods. The French Ministry of Culture has awarded the garden the distinction of *Jardin remarquable*, while the Michelin Green Guide has awarded it three stars.

www.marqueyssac.com +33 5 53 31 36 36

107 PALACE OF VERSAILLES (CHÂTEAU DE VERSAILLES)

Place d'Armes, Versailles, France

TO VISIT BEFORE YOU DIE BECAUSE

These are arguably the most famous palatial gardens in the world.

Try as they might, no royal court in Europe could fully match the splendor of Louis XIV's Château de Versailles, the trendsetter for an entirely new level of monarchical extravagance. While the palace itself is a marvel, the gardens are particularly awe-inspiring. French landscape designer André Le Nôtre perfected the French formal garden here, using perspective, symmetry, and order to assert the king's dominance over nature, and, by proxy, the world. The nearly 2,000-acre gardens were built over the course of 40 years, during the second half of the 17th century, and today they include more than 350,000 trees, 700 topiaries, 50 fountains (though originally there were 2,400), and 221 works of art. After you stroll the Orangery, the parterres, and perhaps even the entire Grand Canal, head to the gardens of the Grand Trianon and the Petit Trianon, the latter of which was the work of Marie Antoinette. The queen brought in English and Asian styles to the gardens, as well as a rustic recreation of a hamlet.

www.chateauversailles.fr +33 1 30 83 78 00

108 AKUREYRI BOTANICAL GARDEN (LYSTIGARÐUR AKUREYRAR)

Eyrarlandsvegur, Akureyri, Iceland

TO VISIT BEFORE YOU DIE BECAUSE

You might not expect plants to thrive just 31 miles south of the Arctic Circle, but the Icelandic city of Akureyri proves that notion wrong.

In 1912, a group of Icelandic women established their country's first public park, located in the northern city of Akureyri. Its botanical gardens, however, weren't planted until much later—in 1957. The Akureyri Botanical Garden was born out of the collection of gardener Jón Rögnvaldsson (1895–1972), who sought to collect not only all of Iceland's flora, but also plants from around the world that might thrive in Akureyri. (Despite being just a few degrees below the Arctic Circle, the city has a surprisingly temperate microclimate.) Today the garden is home to 430 native species and an impressive 6,600 foreign species, as well as a collection of historic wooden buildings, some of which are among the oldest in the city.

www.lystigardur.akureyri.is

+354 462 7487

109 POWERSCOURT ESTATE

Enniskerry, County Wicklow, Ireland

TO VISIT BEFORE YOU DIE BECAUSE

The waterfall is one of the most beautiful in Ireland.

With a history dating back to the 13th century, Powerscourt Estate has undergone quite a few major transformations. The first took place in the 18th century, when the medieval castle was remodeled into a more modern abode, taking on its Palladian style. In the 19th century, Mervyn Wingfield (1836–1904), 7th Viscount Powerscourt, developed the gardens further, adding attractions such as the Japanese Gardens, the Italian Garden, and the Tower Valley. Over the years, the main house and its gardens have been used as a filming location for various films and TV shows, including Stanley Kubrick's *Barry Lyndon* and Kevin Reynolds' *The Count of Monte Cristo*. Today, the property functions as a resort, with an on-site hotel, a golf course, and shops in the old country house. But the best part of staying here overnight is the chance to explore the 47 acres of gardens, including the scenic Powerscourt Waterfall.

www.powerscourt.com +353 1 204 6000

110 ISOLA BELLA

Lake Maggiore, Italy

TO VISIT BEFORE YOU DIE BECAUSE

The terraced Baroque gardens are designed to look like a ship sailing the waters of Lake Maggiore.

Once nothing more than a craggy rock on Lake Maggiore, Isola Bella, one of the Borromean Islands, was transformed into a paradisiacal island with an elaborate palace and is one of the best Baroque gardens in Italy. Its transformation began in the 17th century thanks to the aristocratic Borromeo family, who inaugurated the gardens in 1671, but the architecture and landscape seen today were largely realized throughout the 18th and 19th centuries. The gardens sit on 10 terraces designed to look like a ship from the lake, and they're centered on the grand Teatro Massimo, atop which is a monumental statue of a unicorn, the Borromeos' heraldic symbol. If visiting the island, don't miss the English-style gardens of Isola Bella's neighbor, Isola Madre, which are also spectacular.

www.isoleborromee.it/en/isola-bella +39 323 933478

111 SACRO BOSCO

Giardino (no street number), Bomarzo, Italy

TO VISIT BEFORE YOU DIE BECAUSE

These Mannerist gardens were the subject of a short film by Salvador Dalí.

Of all the 16th-century gardens in Italy, none can compare to Sacro Bosco in the Gardens of Bomarzo. Eschewing typical Renaissance ideals of symmetry and order, military leader and patron of the arts Pier Francesco "Vicino" Orsini (1523–1583), along with architect Pirro Ligorio (c. 1512–1583), designed this garden in the highly stylized Mannerist mode. He filled his emotionally driven garden with grotesque sculptures of mythical monsters and beings like Orcus, a Roman god of the underworld, and of real-life beasts, including one of Hannibal's war elephant. What inspired Orsini? No one knows for sure. But the enigmatic garden has captivated artists for centuries, including Surrealist Salvador Dalí, who made a short film there and incorporated some of its statues into his painting *The Temptation of Saint Anthony*.

www.sacrobosco.it +39 761 924029

112 VILLA D'ESTE, LAKE COMO

Via Regina, 40, Cernobbio, Italy

TO VISIT BEFORE YOU DIE BECAUSE

Is there anything more romantic than a Renaissance garden on Lake Como?

As with many of the great gardens of the world, the one at Villa d'Este on Lake Como (which is not linked to the Villa d'Este, Tivoli, near Rome), began its life in private hands. The estate was built for Cardinal Tolomeo Gallio (1527–1607), whose family owned the property, including its 25-acre gardens, for generations. It changed hands between several illustrious owners before eventually opening as a luxury hotel in 1873, where it attracted all manner of aristocratic clientele. The gardens are still only accesible to hotel guests and known for their magnificent nymphaeum, covered in colorful mosaics, as well as its *Hercules* statue at the top of a hill. Though the main styles of the gardens are Italian Renaissance and Baroque, one of its previous owners, Caroline of Brunswick (1768–1821), did add some English-style landscaping to the grounds in the early 19th century.

www.villadeste.com +39 31 3481

113 VILLA LA MASSA, FLORENCE

Via della Massa, 24 - Florence, Candeli, Italy

TO VISIT BEFORE YOU DIE BECAUSE

You'll feel like you're in the Tuscan countryside at its best just 15 minutes outside of Florence.

Olive trees, irises, lemon groves, cypresses —the 25-acre gardens of Villa La Massa have all the quintessential features of a garden in the Tuscan countryside, and it's only 8 miles outside of Florence. Built in the 13th century, it was then residence for a member of the Medici family, and sold to a series of aristocrats between 1788 and 1948 then converted into a hotel. It's part of the Villa d'Este Hotels since 1998. Guests can now stroll the picturesque Iris Garden on the bank of the Arno or sit beneath lemon trees at the pool with view on the Chianti hills. Dinner will almost certainly include olive oil from the property's own trees.

www.villalamassa.com +39 55 626 11

114 JARDIN EXOTIQUE DE MONACO

62, Boulevard du Jardin Exotique, Monaco

TO VISIT BEFORE YOU DIE BECAUSE

The views from its cliffside location are unparalleled—and it's built on top of a cave that you can tour.

The tiny nation of Monaco might be famous for its Monte Carlo Casino and Formula 1 Grand Prix race, but one of its other highlights is the Jardin Exotique de Monaco. Head to the top of the seaside cliffs overlooking the harbor to find this garden, which offers one of the best views in the entire country. Opened in 1933 by Prince Louis II, the garden houses an expansive collection of more than 1,000 species of succulents from around the world, including cacti and agaves from North America. Some of the plants here are more than 100 years old! But the most surprising part of the garden has little to do with flora; lower down the cliffs is a large cave with evidence of prehistoric human inhabitants. You can tour the cave with a guide, included in the price of admission to the gardens, but beware—you'll want to bring good shoes for the *many* stairs you'll be climbing.

www.jardin-exotique.mc

+377 93 15 29 80

115 HORTUS BOTANICUS LEIDEN

Rapenburg 73, Leiden, the Netherlands

TO VISIT BEFORE YOU DIE BECAUSE

Established in 1590, this inner-city green space is one of the oldest botanical gardens in the world, founded by the man largely responsible for the Netherlands' love for tulips.

When the mayor of Leiden granted Leiden University permission to develop a garden for its medical students in 1590—the first such garden in the Netherlands—the institution put botanist Carolus Clusius (1526–1609) in charge. He tapped into the broad network of the Dutch East India Company to have both live and dried flora sent to the garden from around the world, culminating in an early collection of 1,000 species of plants. His most famous project? Cultivating the tulip, which had not yet been introduced to the Netherlands beyond the walls of his garden. Needless to say, his experiments were successful: the tulip is now inextricably linked to the country. In the centuries to follow, Clusius' successors would continue to grow the collection of tropical subtropical flora, housing them in the Orangery built in 1744. Then, in the early 19th century, the garden received a collection of plants from Japan, courtesy of German physician Philipp Franz von Siebold (1796–1866). Today the 60,000-specimen-strong collection holds such varied plants as the giant water lily, the corpse flower, and the jade vine.

www.hortusleiden.nl +31 71 527 5144

116 KASTEELTUINEN ARCEN

Lingsforterweg 26, Arcen, the Netherlands

TO VISIT BEFORE YOU DIE BECAUSE

These under-the-radar gardens in the grounds of Arcen Castle span more than 100 acres.

As far as palatial country homes go, the 17th-century Arcen Castle in the Netherlands is beautiful, but somewhat modest in scale. Its gardens, however, are extraordinarily monumental. They're larger than their more famous compatriot, Keukenhof—more than 100 acres compared with the flower garden's 79—but draw far fewer crowds, providing visitors a calmer, more leisurely experience. While the gardens nearest to the castle are very traditionally formal, like the elegant Rosarium, which has nearly 8,000 rose bushes, those farther out on the estate take on more different qualities. There are more naturalistic gardens, like the Lommerrijk "leafy" glen, as well as geographical gardens, like the Oosterse Watertuin, designed to look like a Thai fishing village. There's also a contemporary sculpture garden, the Casa Verde greenhouse, and even a mini-golf course!

www.kasteeltuinen.nl +31 77 4736 010

117 KEUKENHOF

Stationsweg 166A, Lisse, the Netherlands

TO VISIT
BEFORE YOU DIE
BECAUSE

This is one of the largest flower gardens in the world, with more than seven million bulbs planted each year.

If you love flowers, you'll love Keukenhof, where more than seven million flowers are planted annually. The Dutch garden got its start as a simple kitchen garden for the 15th-century Teylingen Castle, then was expanded after the construction of Keukenhof Castle in 1641. But the English-style gardens that exist today weren't developed until 1857, when landscape architect Jan David Zocher (1791–1870) and his son Louis Paul Zocher (1820–1915) renovated the grounds. As for the seven million flowers, that tradition didn't begin until 1949, when the grounds were used by bulb growers to exhibit their wares during a springtime exhibition. That yearly event was opened to the public the following year, and it has continued into the present day as a major tourist draw, with an average of 26,000 visitors per day. The gardens are open to the public only eight weeks a year, when the flowers are in bloom, though you can book private events on the grounds throughout the year.

www.keukenhof.nl +31 252 465 555

118 TROMSØ ARCTIC–ALPINE BOTANIC GARDEN

Stakkevollvegen 200, Tromsø, Norway

TO VISIT BEFORE YOU DIE BECAUSE

This is the northernmost botanical garden in the world.

In Tromsø, Norway, the sun never rises above the horizon between November and January during what's known as the polar night. That lack of daylight, however, hasn't prevented the Arctic–Alpine Botanic Garden from thriving. The climate at the world's northernmost botanical garden is surprisingly mild, despite the fact that Tromsø falls just south of the 70th parallel north—the average low temperatures never fall below 20°F, while summer highs reach the low 60s. Therefore, the garden's collection includes Arctic, Antarctic, and alpine plants from all seven continents, including Himalayan blue poppies (*Meconopsis*), as well as naturalistic lichen-covered rock landscapes. Part of the UiT The Arctic University of Norway, the garden is free and open 24 hours a day, 365 days a year—take advantage of that summer sun while it lasts!

www.uit.no/tmu/botanisk +47 776 45 001

119 BACALHÔA BUDDHA EDEN

Quinta dos Loridos, Carvalhal, Portugal

TO VISIT BEFORE YOU DIE BECAUSE

This peaceful Eastern-inspired garden is filled with massive sculptures—and it has wine tastings.

When the Taliban demolished Afghanistan's monumental Bamiyan Buddha statues in 2001, businessman and art patron José "Joe" Berardo (b. 1944) was distraught. The act of "cultural barbarity," as the garden deems it on its website, motivated him to buy Buddha sculptures from across Asia and create a space for them in his native Portugal—more specifically, in the grounds of his winery Bacalhôa. Thus, the Bacalhôa Buddha Eden garden was born. But the 86-acre site is filled with far more than Buddha sculptures. There are 600 hand-painted replicas of China's famous terracotta soldiers, a modern and contemporary sculpture garden with works by such artists as Fernando Botero and Alexander Calder, and an African sculpture garden dedicated to Zimbabwe's Shona people. And, of course, there are plants, from ancient olive trees, to towering palms, to a bamboo forest. Don't miss the wine tastings, either.

www.bacalhoa.pt/enoturismo/bacalhoa-buddha-eden

+351 262 605 240

120 MADEIRA BOTANICAL GARDEN

Caminho do Meio, 9060-345 Funchal, Madeira, Portugal

TO VISIT BEFORE YOU DIE BECAUSE

The panoramic view from the Choreographed Garden is gorgeous.

Ascend via cable car from Monte Parish to the Madeira Botanical Garden, set on the crest of a steep hillside, for both beautiful botanical displays and views over the city of Funchal. Once the private park of the 19th-century Bom Sucesso Estate, the home of the local Reid hospitality family, the 20-acre garden opened to the public in the 1960s and has more than 3,000 species of plants, including a collection dedicated to the endemic flora of the lush Madeira archipelago. Its most famous spot, however, is a flower garden known as the Choreographed Garden, where brightly colored blooms form a geometric pattern over the landscape. There's also a small natural history museum on site filled with fossils and taxidermy.

www.ifcn.madeira.gov.pt/quintas-e-jardins/jardin-botanico-da-madeira-eng-rui-vieira +351 291 211 200

121 MONTE PALACE TROPICAL GARDEN

Caminho do Monte, 174, Funchal, Madeira, Portugal

TO VISIT BEFORE YOU DIE BECAUSE

This is not only a garden, but also a museum of tilework.

Given that Madeira is a botanical oasis, it's not at all surprising that it has more than a few dazzling gardens. Another must-visit spot is the Monte Palace Tropical Garden, a former private estate that had a second life as a luxury hotel. When the hotel closed, its owners sold the property to Portuguese businessman José Berardo in 1987; in its third iteration, it's not only a public garden with an impressive collection of South African cycads, but also a museum with a special focus on the art of tilework, displaying examples from across the globe, from the 15th to the 20th centuries. (A highlight is the *Adventures of the Portuguese in Japan* mural composed of 166 terracotta tiles.) The museum also puts on temporary exhibitions each year.

www.montepalacemadeira.com +351 291 780 800

122 SERRALVES PARK

Rua D. João de Castro, 210, Porto, Portugal

TO VISIT BEFORE YOU DIE BECAUSE

If you've seen enough Renaissance and Baroque gardens, head to the Modernist Serralves for a breath of fresh air.

When Carlos Alberto Cabral, the 2nd Count of Vizela (1895-1968), inherited his family's summer estate in 1923, he had grand visions for modernizing the Victorian-era property. The result was the grand Art Deco–style Casa de Serralves, designed by French architect Charles Siclis (1889-1944) and Portuguese architect José Marques da Silva (1869-1947), and the accompanying 44-acre garden. French landscape architect Jacques Gréber (1882–1962) was in charge of the parkland, and he combined elements of French formal gardens—like the axial plan—with modern, streamlined design. In 1996, the property was reestablished as the Serralves Museum of Contemporary Art and the Serralves Park, the latter of which was updated by landscape architects João Gomes da Silva (b. 1962) and Erika Skabar (b. 1966). Today the park is filled with 8,000 examples of 230 species of endemic and non-native plants, including yew, giant sequoias, and camellias.

www.serralves.pt +351 808 200 543

123 ARBORETUM VOLČJI POTOK

Volčji Potok 3, 1235, Radomlje, Slovenia

TO VISIT BEFORE YOU DIE BECAUSE

This is Slovenia's most visited botanical garden—it has more than 1,000 species of roses and an incredible 2 million tulips.

Once part of a private estate, the Arboretum Volčji Potok is now the most visited botanic garden in the entire country. While the estate dates back to the 17th century, the garden was established by businessman Ferdinand Souvan (1840–1915), who bought the property in the late 19th century, renovated the historic mansion (which burned down in 1944), and planted acres of gardens. His son, Leon Souvan (1877–1949), however, was the creative genius behind the transformation of the 30-acre gardens into a celebrated cultural and natural heritage site in Slovenia. In 1952, several years after the junior Souvan's death, the Arboretum Volčji Potok was established to protect his hard work. Over the decades, the garden has expanded to encompass 210 acres, and it's renowned for its spring tulip exhibition, which features more than 2 million flowers, as well as its two rose gardens planted with a total of more than 1,000 species.

www.arboretum.si +386 1 831 23 45

124 ALHAMBRA Y GENERALIFE

Calle Real de la Alhambra (no street number), Granada, Spain

TO VISIT BEFORE YOU DIE BECAUSE

The architecture, history, and overall grandeur of these iconic Moorish gardens cannot be overstated.

The grand Alhambra complex in Granada, Spain, has a history so long and so monumental that it would take this entire book to tell its story. In a nutshell, today's UNESCO World Heritage site was an ancient fortress that reached its pinnacle under the Muslim Nasrid Dynasty between the 13th and 15th centuries, when it was expanded to become a stand-alone city of Moorish palaces and gardens. While gardens are interwoven throughout the Alhambra complex itself, the gardens of Generalife, a summer palace set just outside the fortress walls, is a real showstopper. Though they've evolved over the years (particularly during the changeovers of empires and during major 20th-century restorations), those gardens are some of the oldest surviving Moorish gardens, and some of the most beautiful.

www.alhambra-patronato.es +34 958 027 971

125 GOTHENBURG BOTANICAL GARDEN

Carl Skottsbergs Gata 22A, Gothenburg, Sweden

TO VISIT BEFORE YOU DIE BECAUSE

This garden is home to Sweden's largest collection of orchids, the extremely rare toromiro tree from Easter Island, and a Michelin-starred rock garden.

Though the Gothenburg Botanical Garden was inaugurated in 1923 during the celebrations of Gothenburg's 300th anniversary, its roots are even older. Botanist and explorer Carl Skottsberg (1880–1963), one of the founders of the garden and its first director, sailed on numerous global expeditions to collect rare plants in the early years of the 20th century. That's one of the reasons the Gothenburg Botanical Garden has an extensive collection of foreign species today, including more than 1,000 species of tropical orchids. One of its rarest plants, however, is the *Sophora toromiro*, or the Easter Island Tree. The tree grew only on the island of Rapa Nui, but it was harvested to extinction. Fortunately, famed Norwegian ethnologist Thor Heyerdahl (1914–2002) preserved seeds from his visit to the island; the tree at Gothenburg grew from one of these seeds. The garden is also home to a highly lauded rock garden, which was awarded two stars by the Michelin Green Guide (the Green Guide is for attractions, which is separate from the restaurant ratings).

www.botaniska.se +46 10-473 77 77

126 DOLMABAHÇE PALACE

Beşiktaş, Istanbul, Turkey

TO VISIT BEFORE YOU DIE BECAUSE

The waterfront gardens belong to the largest palace in Turkey.

Dolmabahçe Palace was born out of a garden in 1856—in fact, its name translates to "filled garden." The palace, Turkey's largest at 11.1 acres of interior space, was constructed on the site of a former royal garden that was created by filling in the Bosphorus. Because the structure is so large, there isn't much real estate for gardens, but what they lack in size, they make up for in style. The entire complex was built in an eclectic aesthetic, which is also reflected in the landscape. The Hasbahçe Garden, for instance, has a more European look with a symmetrical plan, though it is filled with flora from other continents. But the Kuşluk Garden, or bird garden, is more of a natural-looking garden, which is more typical of Turkish gardens of the day. All four of Dolmabahçe's gardens, however, are centered on a pool.

www.millisaraylar.gov.tr/en/palaces/dolmabahce-palace

+90 212 236 90 00

127 EMIRGAN PARK TULIP GARDENS

Emirgan Korusu İçi Yolu No:12, 34467 Sarıyer, Istanbul, Turkey

TO VISIT BEFORE YOU DIE BECAUSE

Thousands of tulips blossom here each spring—it's the hub of Istanbul's annual tulip festival.

Though no one would fault you for associating tulips with the Netherlands, it's not the only country that's enthralled by the cheerful flower. Turkey also has a long history with tulips, which originally grew wild in Central Asia. They were so highly prized by the Ottoman Empire in the middle of the 18th century—sultans wore a tulip in their turbans because they symbolized wealth—that there's even a period named after it. Today, you'll find tulips throughout Turkey, particularly in Istanbul, where there's a major tulip festival each April. One of the best spots to enjoy the blooms is Emirgran Park's tulip gardens, which were planted in the 1960s to help the city revitalize interest in tulip cultivation.

No official website

128 TOPKAPI PALACE

Cankurtaran, Istanbul, Turkey

TO VISIT
BEFORE YOU DIE
BECAUSE

The courtyard gardens provide a peek inside the private lives of Ottoman sultans.

Construction on the remarkable Topkapi Palace in Istanbul, the longtime home of Ottomon sultans and the hub of the government, began in 1460 under Fatih Sultan Mehmet, with major additions built throughout the 16th century and minor ones added through the 19th century. Unlike many palace gardens, which spread outward from the main building, Topkapi's best gardens are in its interior courtyards, which are filled with centuries-old trees, hundreds of thousands of plants, fountains, pools, and pavilions, and often surrounded by elegant architectural features like arched porticos. Botany was often a particular interest for sultans, who cultivated flowers in the gardens—today, tulip gardens fill the courtyards in a nod to the 18th-century Tulip Era of the Ottoman Empire, during which the nation was swept up in a craze over the colorful plant.

www.millisaraylar.gov.tr/en/palaces/
topkapi-palace

+90 212 256 90 00

129 BOTANIC GARDENS (BELFAST)

College Park, Botanic Avenue, Belfast, Northern Ireland, United Kingdom

TO VISIT BEFORE YOU DIE BECAUSE

Its gorgeous Victorian-era conservatory predates the greenhouses at Kew.

Founded as a members-only garden by the Belfast Botanic and Horticultural Society in 1828 (the public was permitted to visit only on Sundays), the Botanic Gardens in Belfast formally became public in 1895. Since the beginning, the garden has maintained a vast collection of warm-climate flora from the Southern Hemisphere, so it was necessary to build a greenhouse from the start. Architect Sir Charles Lanyon (1813–1889) designed the magnificent Victorian-style Palm House, which opened in 1840 and is one of the oldest examples of curved iron-and-glass conservatories. In fact, the ironwork was completed by Richard Turner (1798–1881), who would later go on to build the iconic Palm House at the Royal Botanic Gardens, Kew, in London. The gardens also feature the Tropical Ravine, another greenhouse in which a sunken glen filled with tropical plants, from orchids to banana trees, can be viewed from a balcony.

www.belfastcity.gov.uk/things-to-do/parks-and-open-spaces/a-z-parks/botanic-gardens + 44 28 9031 4762

130 CULZEAN CASTLE

Maybole, South Ayrshire, Scotland, United Kingdom

TO VISIT BEFORE YOU DIE BECAUSE

The castle has one of the largest walled gardens in Scotland—and fabulous ocean views.

At Culzean Castle, the gardens are all about balance—there are both formal spaces with immaculately manicured shrubbery and "wild" areas where nature is largely left to its own devices. The castle itself was built in the 18th century by David Kennedy, the 10th Earl of Cassillis (?–1792), who hired Scottish architect Robert Adam (1728–1792) to design a country house that would flaunt his wealth. (It worked, attracting such noted guests as U.S. President Eisenhower, who was ultimately gifted an apartment in the house.) The setting was 600 acres on the Ayrshire cliffs, which provided plenty of space for all sorts of gardens. Next to the house are the formal gardens and their grand fountain, an orangery, and one of the largest walled gardens in Scotland, where gardeners have been cultivating produce such as peaches and onions for nearly two centuries. Beyond that, the grounds take on a more relaxed atmosphere—today the woods, beaches, and ponds are part of a country park open to the public.

www.nts.org.uk/visit/places/culzean +44 1655 884455

131 HAMPTON COURT GARDENS

East Molesey, Surrey, England, United Kingdom

TO VISIT BEFORE YOU DIE BECAUSE

The 60 acres of formal gardens at Hampton Court Palace are home to the world's oldest surviving hedge maze and the largest grapevine in the world.

With a 500-year history, it's no surprise that the gardens of Hampton Court Palace are some of the most interesting in the world. The palace itself was built in the early 16th century, originally for one of the ministers of King Henry VIII (1491–1547), but it was eventually given to the monarch—today it belongs to Queen Elizabeth II (b. 1926). The first small gardens were planted before Henry VIII moved in, while the king expanded the grounds to include his Privvy Garden and a park for deer hunting (the descendants of those deer still live on the grounds today). Currently, the formal gardens cover 60 acres, while there are some 750 acres of parkland. Highlights include the world's oldest surviving hedge maze, which was commissioned by William III around 1700, as well as the world's largest grapevine, planted by Lancelot "Capability" Brown in 1768—its base has a circumference of 13 feet, and the longest rod reaches 120 feet in length.

www.hrp.org.uk/hampton-court-palace/whats-on/hampton-court-gardens

+44 333 320 6000

132 KENSINGTON PALACE

Kensington Gardens, London, England, United Kingdom

TO VISIT BEFORE YOU DIE BECAUSE

John Evelyn, an 18th-century writer, described the gardens as "very delicious."

Originally a "modest mansion," per the Historic Royal Palaces, the Kensington Palace Royal Residence was expanded upon by acclaimed English architect Sir Christopher Wren (1632–1723) at the behest of King William II (1650–1702) and Queen Mary II (1662–1694), who purchased the property in 1689. It's still occupied by royals today: the Duke and Duchess of Cambridge, better known as Prince William (b. 1982) and Kate Middleton (b. 1982), live there with their children. Under William II and Mary II, stately gardens were planted around the palace, but they were greatly expanded upon by the next occupants. Mary's sister, Queen Anne (1665–1714), built the Orangery in 1704, while Queen Caroline (1683–1737) developed the now-public park, including the Serpentine lake. Today, the palace gardens include the Sunken Garden, built in 1908; a wildflower meadow; Queen Anne's Orangery; and the original formal gardens, which were described by writer John Evelyn as "very delicious."

www.hrp.org.uk/kensington-palace/whats-on/the-palace-gardens

+44 20 3166 6115

133 LOST GARDENS OF HELIGAN

Pentewan, St. Austell, Cornwall, England, United Kingdom

TO VISIT BEFORE YOU DIE BECAUSE

These mysterious gardens were once lost to the history books—until someone found a hidden door leading to them.

The story of the Lost Gardens of Heligan reads like a dramatic and tragic work by a novelist—say, *The Secret Garden* by Frances Hodgson Burnett (1849–1924). But this is no fictional tale. The 13th-century Heligan manor was purchased by the Tremayne family in the 16th century, and it and its remarkable gardens flourished through the 20th century. But World War I whisked most of the estate's 22 gardeners to battle, and many of them did not survive. Without its keepers, the garden became overgrown, abandoned, and eventually forgotten. Then, in 1990, a distant Treymane family member, John Willis, and businessman Tim Smit (b. 1954) discovered a door in a derelict building on the grounds of the estate—through it were the Lost Gardens of Heligan. A major restoration project has since resurrected the gardens, and their 200 acres are now open for the public to explore.

www.heligan.com +44 1726 845100

134 QUEEN MARY'S GARDENS

The Regent's Park, London, England, United Kingdom

TO VISIT BEFORE YOU DIE BECAUSE

The rose garden and its 12,000 roses have a claim to fame in film and literature.

Fans of Disney's *101 Dalmatians* animated film and fans of writer Sylvia Plath both might recognize Queen Mary's Gardens, the first from the "meet cute" scene between the human and canine main characters, and the second from the poem of the same name. That simply goes to show how varied audiences can be captivated by its blooms. The rose garden might be the most famous part of this section of The Regent's Park—it's no surprise, with its 12,000 individual plants that bloom each May—but Queen Mary's Gardens is also home to beds of begonias and delphiniums, as well as a Mediterranean garden. These plantings, however, are a relatively recent addition to the site. Queen Mary's Gardens were formally established in 1932, but the earliest records of the grounds date back to 1086, when they were part of a medieval manor.

www.royalparks.org.uk/parks/the-regents-park/things-to-see-and-do/gardens-and-landscapes/queen-marys-gardens

+44 300 061 2300

135 ROYAL BOTANIC GARDENS, KEW

Kew, Richmond, London, England, United Kingdom

TO VISIT BEFORE YOU DIE BECAUSE

This UNESCO World Heritage site has the world's largest collection of living plants, growing on spectacular grounds with historic architecture.

If it grows on this earth, you'll find it at the Royal Botanic Gardens in Kew. That might be a bit of an exaggeration—but not much of one. Kew Gardens, as it's more commonly known, holds the largest collection of living plants in the world, with more than 50,000 plants growing on its sprawling grounds, plus millions of preserved specimens within its research centers. It was founded, however, as a far more humble 9-acre garden in 1759 by Princess Augusta, mother of King George III. Over its long history, Kew Gardens expanded to 300 acres at the main site in London, plus a separate 500-acre wild garden in Sussex called Wakehurst. It was named a UNESCO World Heritage site in 2003, not only for its flora, but also for its historic architecture, from the iconic glass Palm House to the towering Great Pagoda. Kew Gardens is also home to the Millennium Seed Bank, which holds more than 2.25 billion seeds from around the world.

www.kew.org +44 20 8332 5655

136 SKY GARDEN

1 Sky Garden Walk, London, England, United Kingdom

TO VISIT BEFORE YOU DIE BECAUSE

Most gardens are on the ground, but this one is in the clouds.

As its name implies, the Sky Garden in London isn't your standard ground-bound garden; it's located on the 35th through the 37th floors of the famous "Walkie Talkie" skyscraper by Uruguayan architect Rafael Viñoly, formally called 20 Fenchurch Street, making it the city's highest public garden. Naturally, it offers 360-degree views of London. As the gardens by landscape firm Gillespies are indoors in a three-story climate-controlled atrium—though there is an outdoor deck—colorful Mediterranean and South African plants like the bird of paradise and French lavender fill the interior terraces. Though it's no Kew Gardens, it does have a pretty spectacular vantage point from which to observe London. The garden is free to the public, but you must book a timed ticket in advance of your visit, or you can make a reservation at one of the gardens' restaurants and bars. Time your visit to sunrise or sunset for the best light!

www.skygarden.london +44 20 7337 2344

137 ADELAIDE BOTANIC GARDEN

North Terrace, Adelaide, South Australia, Australia

TO VISIT BEFORE YOU DIE BECAUSE

This garden bookends the history of greenhouse design with the Palm House from 1877 and the Bicentennial Conservatory from 1989.

Opened in 1857, the 124-acre Adelaide Botanic Garden is not only a paradise for plant lovers, but also one for architecture lovers. The gardens feature three distinctive buildings showcasing highly different architectural forms: the Victorian-style Palm House, built in 1877; the Greek-style Santos Museum of Economic Botany, built in 1881; and the contemporary UFO-like Bicentennial Conservatory, built in 1989, which is the Southern Hemisphere's largest single-span conservatory. Plant-wise, the Adelaide Botanic Garden has 11 main outdoor zones, including two rose gardens, the family-friendly educational Little Sprouts Kitchen Garden, and the Australia Native Garden. Inside the greenhouses, highlights include the *Victoria amazonica* water lily and the *Amorphophallus titanum* corpse flower. While inside the Santos Museum, you'll find papier-mâché models of fruit and fungi from the 19th century, as well as a cabinet of curiosities.

www.botanicgardens.sa.gov.au/visit/adelaide-botanic-garden

+61 8 8222 9311

138 BRISBANE BOTANIC GARDENS MOUNT COOT-THA

Mount Coot-tha Road, Toowong, Queensland, Australia

TO VISIT BEFORE YOU DIE BECAUSE

The Tropical Display Dome is a fantastic sight, both inside and out.

Covering an area of nearly 130 acres outside of the city center, the Brisbane Botanic Gardens Mount Coot-tha are a major addition to the central business district's City Botanic Gardens, which didn't have room for expansion and were prone to floods. With so much space, the gardens, which opened to the public in 1976, are able to display a broad variety of flora—more than 200,000 individual specimens. Brisbane Botanic Gardens Mount Coot-tha has the world's largest collection of trees native to the Australian rain forest, as well as an impressive collection of other native Australian plants. In the Japanese Garden, Australian and nonnative plants are combined in a plan by master landscape architect Kenzo Ogata (1912–1988) that adheres to classic Japanese design principles; next door is the Bonsai House. Outstanding architectural features of the garden are the geodesic Tropical Display Dome, opened in 1977, and the Sir Thomas Brisbane Planetarium, opened a year later.

www.brisbane.qld.gov.au/things-to-see-and-do/council-venues-and-precincts/parks/botanic-gardens-in-brisbane/brisbane-botanic-gardens-mt-coot-tha

+61 7 3403 2535

139 CHINESE GARDEN OF FRIENDSHIP

Pier Street, Cnr Harbour Street, Darling Harbour, Sydney, New South Wales, Australia

TO VISIT BEFORE YOU DIE BECAUSE

The Ming-style garden is a grand display of Taoist principles—and an oasis for people in Sydney's central business district.

To celebrate a deep friendship between sister cities Sydney, Australia, and Guangzhou, China, landscape designers from the latter developed this small-scale version of a classic Ming-era garden, which was built in Sydney's central business district by Australian craftspeople. The 2.5-acre garden follows the traditional philosophical principles of Taoism: yin-yang, symbolizing balanced opposing forces, and wuxing, representing the five elements (fire, water, wood, metal, and earth). These principles lend themselves to a more naturalistic garden than the formal styles of the Western world, with recreations of landscape elements like waterfalls and forests. Throughout the Chinese Garden of Friendship are pavilions, rock sculptures, granite bridges, and the Dragon Wall, a glazed-tile screen depicting one brown and one blue dragon, representing the sister states of Guangdong and New South Wales, respectively.

www.darlingharbour.com/precincts/chinese-garden

+61 2 9240 8888

140 ROYAL BOTANIC GARDEN SYDNEY

Mrs Macquaries Road, Sydney, New South Wales, Australia

TO VISIT BEFORE YOU DIE BECAUSE

This is the oldest scientific institution in Australia—and it has glorious views of the Sydney Opera House.

When it was founded in 1816, the Royal Botanic Garden Sydney was not only the first botanical garden in Australia, but also the country's first scientific institution. Its first superintendent, Charles Fraser (1788–1831), set off on expeditions across the continent, collecting plants to bring back to Sydney for cultivation and study. Today the gardens are massive—74 acres, to be exact—and include the Australian Rainforest Garden and the Australia Native Rockery, both home to the country's native species, as well as a rose garden, a palm grove, and an Asiatic Garden, which houses international species. Don't miss your chance to see the extremely rare Wollemi pine, thought to be extinct until it was discovered in a single canyon system in 1994. The Royal Botanic Garden Sydney has established the Wollemi Pine Conservation Program to study and protect the ancient "dinosaur tree."

www.rbgsyd.nsw.gov.au +61 2 9231 8111

141 ROYAL BOTANIC GARDENS VICTORIA – CRANBOURNE GARDENS

Ballarto Road and Botanic Drive, Cranbourne, Victoria, Australia

TO VISIT BEFORE YOU DIE BECAUSE

Though you'll be just 45 minutes outside Melbourne's city center, you'll feel as if you're in the middle of the bush.

The Cranbourne division of the Royal Botanic Gardens Victoria was completed in 2012 after years of development. It includes nearly 900 acres of protected native bushland filled with miles of winding trails for walking and biking. Some 37 of those acres are landscaped gardens that specialize in the flora of Australia; the focal point of this part of the grounds is the Red Sand Garden, designed to recall the continent's Red Centre. All together, there are 170,000 plants in the garden, representing 1,700 species. Cranbourne's sister garden, the other half of the Royal Botanic Gardens Victoria, was founded in 1846 in Melbourne, and its collection focuses on global species.

www.rbg.vic.gov.au/cranbourne-gardens +61 3 5990 2200

142 ROYAL TASMANIAN BOTANICAL GARDENS

Lower Domain Road, Queens Domain, Hobart, Australia

TO VISIT BEFORE YOU DIE BECAUSE

Tasmania's only botanical garden is home to the world's only subantarctic plant house.

Tasmania is already pretty chilly by Australian standards, given its location south of the mainland. But at the Royal Tasmanian Botanical Gardens, you can be transported to an even colder climate—to the subantarctic, to be more specific. The garden has the only subantarctic plant house in the world, and it showcases the flora found on UNESCO World Heritage site Macquarie Island, located about halfway between Tasmania and Antarctica. Of course, that's not all you'll find here. The garden is Australia's second oldest, founded in 1818, and the only botanic garden in Tasmania. It holds 42 distinct collections, including a Japanese garden and a Chinese garden, with more than 6,000 species in all, from significant trees to herbs.

www.rtbg.tas.gov.au

+61 3 6166 0451

143 WENDY WHITELEY'S SECRET GARDEN

Lavender Street, Lavender Bay, Sydney, New South Wales, Australia

TO VISIT BEFORE YOU DIE BECAUSE

Despite its name, this garden is largely considered to be the worst-kept secret in Sydney.

Wendy Whiteley lived a titanic life before she planted her garden. The former wife of Australian artist Brett Whiteley (1939–1992), she traveled the world with her then-husband, living in Europe, the United States, and even Fiji. But the couple ultimately returned to Australia with their daughter, Arkie (1964–2001), settling in Sydney's Lavender Bay. Eventually they divorced, but after both Brett's and Arkie's deaths, Wendy Whiteley channeled her grief into clearing the garbage-filled lot next to her home, ultimately developing what would become Wendy Whiteley's Secret Garden. "I didn't know anything about horticulture when I started the garden. I just knew what I liked," she said. "I've since learnt what likes being here. It's a symbiotic relationship between the plants, myself, and my gardeners." The hillside garden is a peaceful space open to the public, offering tranquility with a lovely view of Sydney Harbour Bridge.

www.wendyssecretgarden.org.au

144 WESTERN AUSTRALIAN BOTANIC GARDEN

Fraser Avenue, Perth, Western Australia, Australia

TO VISIT BEFORE YOU DIE BECAUSE

There's a 750-year-old giant boab tree in this urban park.

Part of the immense, 990-acre Kings Park in Perth, the 42-acre Western Australian Botanic Garden opened in 1965 with the goal of cultivating and displaying plants native to the state, many of which are not found anywhere else in the world. It currently holds about one-quarter of Western Australia's 12,000 species of native plants—visit the Conservation Garden to see rare and endangered plants from the region. The garden is also home to a 750-year-old, 40-ton giant boab tree that was moved to the park from 2,000 miles away, as it was in the way of a construction project. The tree, one of the biggest attractions in the park (literally), is named "Gija Jumulu" in honor of the Gija Indigenous group local to its original home and their name for the species.

www.bgpa.wa.gov.au/kings-park/area/wa-botanic-garden

+61 8 9480 3600

RioTinto

145 GARDEN OF THE SLEEPING GIANT

Wailoko Road, Nadi, Fiji

TO VISIT BEFORE YOU DIE BECAUSE

More than 2,000 types of orchards fill this peaceful garden, which has ties to Old Hollywood.

Raymond Burr (1917–1993) might be best known for his leading role in the TV drama *Perry Mason*, but the American actor was also an amateur horticulturist with a particular passion for orchids—and for Fiji. He combined those two loves by purchasing a private estate in the island nation in 1977, then importing thousands of species of orchids to fill its grounds. Burr's collection, including a number of hybridizations that he cultivated, eventually made its way to the island of Viti Levu, where it now lives a second life as the 49-acre Garden of the Sleeping Giant, so named for the rock formation beneath which it sits. Today the gardens have more than 2,000 species of orchids interwoven with other greenery, including lily ponds and native rain forest.

www.gardenofsleepinggiant.com +679 672 3418

146 AUCKLAND BOTANIC GARDENS

102 Hill Road, Manurewa, Auckland, New Zealand

TO VISIT BEFORE YOU DIE BECAUSE

This young garden already has a collection 10,000 plants strong.

Despite being a relatively young botanic garden—it opened in 1982—Auckland Botanic Gardens impresses with its collection of 10,000 living plants from around the world that draw in more than a million people each year. Beyond dazzling visitors with its displays of native and non-native flora, which spans everything from roses to palm to harakeke (New Zealand flax), the gardens host the Sculpture in the Gardens exhibition every other summer. The show displays works by New Zealand artists, many of which enter the gardens' permanent collection that's installed in the Huakaiwaka Visitor Center and throughout its grounds. The Auckland Botanic Gardens are also an important research and conservation facility, protecting threatened species from all over the world, such as *Camellia nitidissima* from China.

 www.aucklandbotanicgardens.co.nz +64 9 267 1457

147 AYRLIES GARDEN AND WETLANDS

125 Potts Road, Whitford, Auckland, New Zealand

TO VISIT BEFORE YOU DIE BECAUSE

Despite its young age, it's already been named one of the New Zealand Gardens of International Significance.

In 1964, Ayrlies Garden was nothing more than an empty paddock on the Auckland property of Beverley and Malcolm McConnell (1930–1995). Just 50 years later, it's already been recognized as a New Zealand Garden of International Significance. Inspired by the Hawke's Bay country garden of her childhood, Bev, as she's known to family and friends, has expertly combined lawns, trees, flowers, shrubs, ponds, and waterfalls like a painter blending oils; the garden feels natural, borderline wild, and yet at the same time perfectly composed, with an expert focus on color. "I think gardens are so good for the soul," Bev said in a 2012 interview. In 2000, a 35-acre wetland was added to Ayrlies, providing a refuge for wildlife.

www.ayrlies.co.nz +64 9 5308 706

148 CHRISTCHURCH BOTANIC GARDENS

Rolleston Avenue, Christchurch, New Zealand

TO VISIT BEFORE YOU DIE BECAUSE

Christchurch is known as "The Garden City," and its botanic gardens demonstrate why.

Inaugurated in 1863, the Christchurch Botanic Gardens has 10 flowering gardens, one garden dedicated to native New Zealand flora, and six conservatories (including the 1923-built Cunningham House, home to an *Amorphophallus titanum* corpse flower) across its 52-acre grounds—it's no wonder that Christchurch is nicknamed "The Garden City." The gardens are also a hub for contemporary art, thanks to its partnership with SCAPE Public Art, which brings in the work of both established and emerging local and international artists; the home of the New Zealand World Peace Bell; and a historic site important to Antarctic exploration. Both Sir Ernest Shackleton and Sir Robert Falcon Scott used the Magnetic Observatory that once existed here to calibrate their instruments before their expeditions.

www.ccc.govt.nz/parks-and-gardens/christchurch-botanic-gardens

+64 3 941 8999

149 LARNACH CASTLE & GARDENS

145 Camp Road, Dunedin 9077, New Zealand

TO VISIT BEFORE YOU DIE BECAUSE

These gardens are set on the grounds of New Zealand's only "castle."

While it may not be a true, fortified castle, Larnach Castle certainly has the look of one. The Gothic Revival structure, designed by Dunedin architect R. A. Lawson (1833–1902), was built as the private home of businessman and politician William Larnach (1833–1898). Though gardens existed on the original property, the most recent iteration is the work of current owner Margaret Barker (b. 1942), who purchased the castle in 1967 and spent decades restoring and developing the house and the grounds, which now include 7 acres of gardens featuring flower beds, a laburnum tunnel, a rainforest garden with giant lilies, and a rock garden. Throughout the gardens are sculptural nods to *Alice's Adventures in Wonderland*; keep an eye out for the Cheshire Cat!

www.larnachcastle.co.nz +64 3 476 1616

150 OTAHUNA LODGE

224 Rhodes Road, Tai Tapu, New Zealand

TO VISIT BEFORE YOU DIE BECAUSE

The largest historic private residence in New Zealand has 30 acres of gardens, including a daffodil field with millions of bulbs.

Sir Heaton Rhodes (1861–1956) was not only a lawyer and a politician, but also a celebrated horticulturist who served as president of the Canterbury Horticultural Society for most of his life. It's no surprise, then, that his private home, today the grand Otahuna Lodge hotel, has a pretty spectacular garden. Built in 1895, the Queen Anne–style structure is the largest historic private residence in New Zealand, with 30 acres of gardens that include a Dutch garden, a games lawn, and a magnificent field of daffodils—one of Rhodes' favorite flowers—where millions of bulbs bloom each year. Guests at the hotel can also visit the Potager Garden with the property's chef to harvest ingredients for the day's dinner.

www.otahuna.co.nz/gardens +64 3 329 6333

© Photos

p. 12 Paul Vinten – iStock / p. 13 (top) Paul Vinten – iStock / p. 13 (bottom) Dennis Dongelmans – iStock / p. 14 Cyanid / p. 15 Gerhard Petterson – Shutterstock / pp. 16-17 Fela Sanu – iStock / p. 18 Natalia Sterleva – iStock / p. 19 © Albina Bauer / p. 20-21 © Albina Bauer/ p. 22 Courtesy of La Mamounia / p. 23 Rytis Bernotas – iStock / p. 24 lcswart – iStock / p. 25 EcoPic – iStock / pp. 26-27 Ian Murdoch – iStock / p. 28 Rod Waddington – Flickr / p. 29 Johanes Duarte – iStock / p. 30 Agata Fetschenko – iStock / p. 31 saiko3p – iStock/ p. 32 (top) Natalia SO – iStock / p. 32 (bottom) Natalia SO – iStock / p. 33 lisandrotrarbach – iStock / p. 34 Sharon Cooke / p. 35 (top) - Sharon Cooke / p. 35 (bottom) - Sharon Cooke / pp. 36-37 Sharon Cooke / p. 38 Sue Stubbs / p. 39 Courtesy of Jardín Botánico Joaquín Antonio Uribe / p. 40 Roman_Rahm – iStock / p. 41 (top) Hanis - iStock / p. 41 (bottom) zxvisual - iStock / p. 42 wwing- iStock / p. 43 canbalci – iStock / p. 44 gregobagel – iStock / p. 45 benedek – iStock / p. 46 LiamDoerksen – iStock / p. 47 Max Donoso / pp. 48-49 Tomás Fabre / p. 50 Courtesy of Jardín Botánico Joaquín Antonio Uribe / p. 51 pxhidalgo – iStock / p. 52 Christer T Johansson – Wikimedia / p. 53 Carlos Vargas – iStock / p. 54 Roberto Michel – iStock / p. 55 Quasarphoto – iStock / p. 56 Donnie McKinnon / p. 57 Donnie McKinnon / pp. 58-59 Donnie McKinnon / p. 60 Ultima_Gaina – iStock / p. 61 Skyhobo – iStock / p. 62 (top) beklaus – iStock / p. 62 (bottom) WGCPhotography – iStock / p. 63 lightphoto – iStock / p. 64 4FR - iStock / p. 65 iShootPhotosLLC – iStock / pp. 66-67 CampPhoto- iStock / p. 68 (top) Courtesy Fairchild Tropical Botanic Garden - Photo by David Sutta / p. 68 (bottom) Courtesy Fairchild Tropical Botanic Garden - Photo by David Sutta / pp. 70-71 Courtesy Fairchild Tropical Botanic Garden - Photo by David Sutta / p. 72 (top) Courtesy of Fort Worth Botanic Garden / p. 72 (bottom) Courtesy of Fort Worth Botanic Garden / p. 73 Courtesy of Fort Worth Botanic Garden / p. 74 Dmitrii Sakharov – iStock / p. 75 Erik Kvalsvik / pp. 76-77 Helen Norman / p. 78 Tom Hennessy / p. 79 Courtesy of Limahuli Garden and Preserve / p. 80 Courtesy of Longwood Gardens / p. 81 Courtesy of Longwood Gardens / pp. 82-83 Courtesy of Longwood Gardens / p. 84 Kent Burgess - Courtesy of Missouri Botanical Garden / p. 85 Bim - iStock / p. 86 Courtesy of The New York Botanical Garden / p. 87 Courtesy of The New York Botanical Garden / p. 88 Courtesy of Philadelphia's Magic Gardens / p. 89 Courtesy of Philadelphia's Magic Gardens / p. 90 (top) tvphoto – iStock / p. 90 (bottom) rruntsch – iStock / p. 91 photoquest7 – iStock / pp. 92-93 joshuaraineyphotography- iStock / p. 94 Zane Michael Cooper- iStock / p. 95 Courtesy of United States Botanic Garden / pp. 96-97 Courtesy of United States Botanic Garden / p. 98 Robin Hill Photography - Courtesy of Vizcaya Museum and Gardens Lalbagh Fort / p. 99 Fahad Hossain – iStock / p. 100 frentusha – iStock / p. 101 SCQBJ-JZ – iStock / p. 102 axz66 – iStock / p. 103 bpperry – iStock / pp. 104-105 luxizeng – iStock / p. 106 f9photos – iStock / p. 107 Sergdid – iStock / p. 108 CHUNYIP WONG – iStock / p. 109 huad262 – iStock / p. 110 powerofforever – iStock / p. 111 mattwicks – iStock / p. 112 Roop_Dey – iStock / p. 113 SBellinger- iStock / pp. 114-115 Skouatroulio – iStock / p. 116 ugurhan – iStock / p. 117 Konstantin_Novakovic – iStock / p. 118 Cezary Wojtkowski – iStock / p. 119 gyro – iStock / p. 120 Courtesy of Adachi Museum of Art Hōkoku-ji / p. 121 dar_st – iStock / p. 122 Courtesy of Hotel Chinzanso Tokyo / p. 123 Courtesy of Hotel Chinzanso Tokyo / p. 124 JayTurbo – iStock / p. 125 p. 125 Photo by Japanexperterna.se. License: CC BY-SA 3.0 / pp. 126-127 LeeYiuTung – iStock / p. 128 Sean Hsu – Shutterstock / p. 129 Hiro1775 – iStock / p. 130 BmtNob – Shutterstock / p. 131 tupungato – iStock / pp. 132-133 alina-sharoik – Shutterstock / p. 134 sdstockphoto – iStock /p. 135 sdstockphoto – iStock / p. 136 oxico- iStock / p. 137 OSTILL – iStock / p. 138 Timon Schneider- iStock / p. 139 Muhammad Rafi Ullah – iStock / pp. 140-141 Muhammad Rafi Ullah – iStock / p. 142 Vladislav Zolotov – iStock / p. 143 mikolajn – iStock / p. 144 lavendertime – iStock / p. 145 primeimages – iStock / p. 146 KiranPix - Shutterstock / p. 147 Photon-Phothos – iStock / pp. 148-49 - Photon-Photos – iStock / p. 150 GoranQ – iStock / p. 151 Kanawa_Studio – iStock / p. 152 Nuwan Liyanage – iStock / p. 153 yellowhey – iStock / p. 154 Dmitry Malov – iStock / p. 155 huafires – iStock / pp. 156-157 artpritsadee – iStock / p. 158 - dannay79 – iStock / p. 159 Gelia – iStock / pp. 160-161 Gelia – iStock / p. 162 WhiteLacePhotography – iStock / p. 163 alfaori – iStock / p. 164 spastonov – iStock / p. 165 StudioBarcelona – iStock / p. 166 Daderot / p. 167 Flavio Vallenari – iStock / p. 168 mammuth – iStock / p. 169 Vladislav Zolotov – iStock / p. 170 ClarkandCompany – iStock / p. 171 chrisfarrugia – iStock / p. 172 IrinaSen – iStock / p. 173 marcolevallois – iStock / p. 174 © Michel GOGNY-GOUBERT - Normandy Tourist Board Château et les Jardins de Villandry / p. 175 boivin nicolas - Shutterstock / pp. 176-177 boivin nicolas - Shutterstock / p. 178 irakite – iStock / p. 179 Studio-Annika – iStock / p. 180 Thomas321 – iStock / p. 181 Photo by xiquinhosilva. License: CC BY 2.0 / pp. 182-183 Photo by Paolo Costa Baldi. License: GFDL/CC-BY-SA 3.0 / p. 184 Photo by Alexander Grebenkov. License: CC BY 3.0 / p. 185 alexandrumagurean – iStock / p. 186 (top) xenotar – iStock / p. 186 (bottom) precinbe – iStock / p. 187 olgysha2008 – iStock / p. 188 Photogilio – iStock / p. 189 Archives Villa d'Este / pp. 190-191 Archives Villa d'Este / p. 192 Archives Villa La Massa / p. 193 Archives Villa La Massa / p. 194 lkonya – iStock / p. 195 Courtesy of Hortus botanicus Leiden / p. 196 Courtesy of Castle Gardens Arcen - Kasteeltuinen Arcen / p. 197 Olena_Z – iStock / pp. 198-199 nikitje – iStock / p. 200 Kristian Nyvoll / p. 201 Courtesy of Bacalhôa Buddha Eden / pp. 202-203 Courtesy of Bacalhôa Buddha Eden / p. 204 The World Traveller – iStock / p. 205 Alessandro Barone – iStock / p. 206 LuisPortugal iStock / p. 207 Samo Trebizan – iStock / p. 208 Adrian Wojcik – iStock / p. 209 Martin Wahlborg – iStock / pp. 210-211 Martin Wahlborg – iStock / p. 212 Alx_Yago – iStock / p. 213 serts – iStock / p. 214 Serg Zastavkin - Shutterstock / p. 215 benkrut – iStock / p. 216 DouglasMcGilviray – iStock / p. 217 Vladislav Zolotov – iStock / p. 218 SangHyunPaek – iStock / p. 219 (top) chrisdorney – iStock / p. 219 (bottom) lightphoto – iStock / p. 220 Courtesy of the Lost Gardens of Heligan / p. 221 Luke Abrahams – iStock / p. 222 Courtesy of the Royal Botanic Gardens, Kew / p. 223 Courtesy of Sky Garden – rhubarb / pp. 224 – 225 Courtesy of Sky Garden – rhubarb / p. 226 (top) zensu - iStock / p. 226 (bottom) Acerebel – iStock / p. 228 Paul Harrop / p. 229 (top) Phil Hargreave / p. 229 (bottom) Paul Harrop / p. 230 (top) Photon-Photos – iStock / p. 231 (bottom) Photon-Photos – iStock / p. 232 Courtesy of Royal Botanic Garden Sydney / p. 233 Courtesy of Royal Botanic Gardens Victoria / pp. 234-235 Courtesy of Royal Botanic Gardens Victoria / p. 236 Courtesy of Royal Tasmanian Botanical Gardens / p. 237 Courtesy of Royal Tasmanian Botanical Gardens / p. 238 Jason Busch / p. 239 Jason Thomas / pp. 240 – 241 Jason Thomas / p. 242 Courtesy of Tourism Fiji / p. 243 denizunlusu – iStock / p. 244 Bruce Clarke Incredible Images Auckland NZ / p. 245 Bruce Clarke Incredible Images Auckland NZ / pp. 245-246 Bruce Clarke Incredible Images Auckland NZ / p. 248 cmfotoworks – iStock / p. 249 Photo by Bernard Spragg. License: CC0 1.0 / p. 250 Courtesy of Otahuna Lodge / p. 251 Courtesy of Otahuna Lodge /pp. 252-253 Courtesy of Otahuna Lodge

In the same series

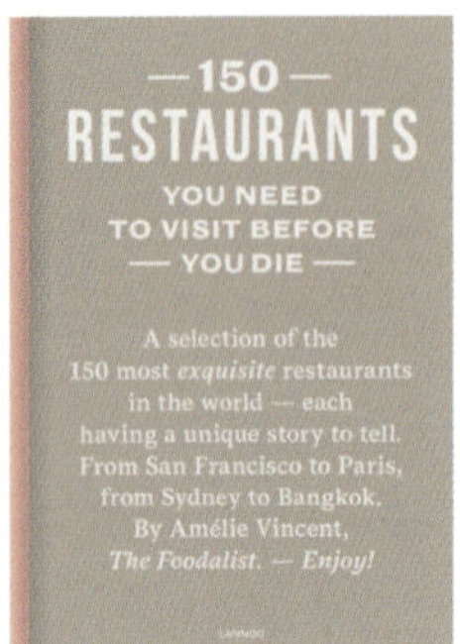

150 Restaurants
You Need to Visit
Before You Die
ISBN 9789401454421

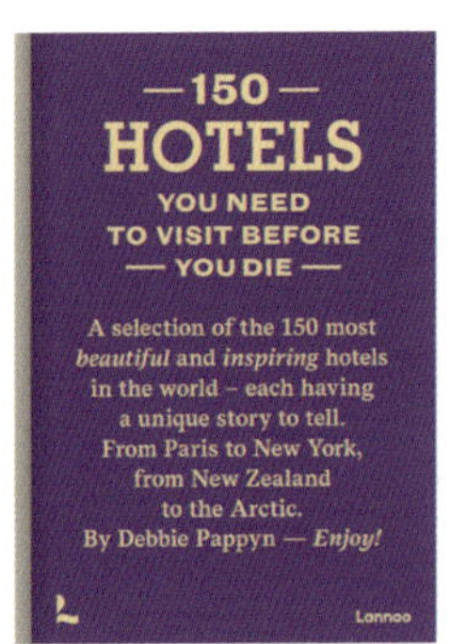

150 Hotels
You Need to Visit
Before You Die
ISBN 9789401458061

150 Golf Courses
You Need to Visit
Before You Die
ISBN 9789401481953

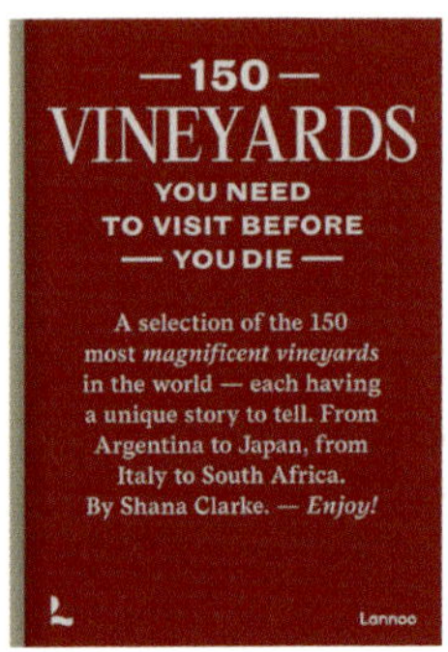

150 Vineyards
You Need to Visit
Before You Die
ISBN 9789401485463

150 Houses
You Need to Visit
Before You Die
ISBN 9789401462044

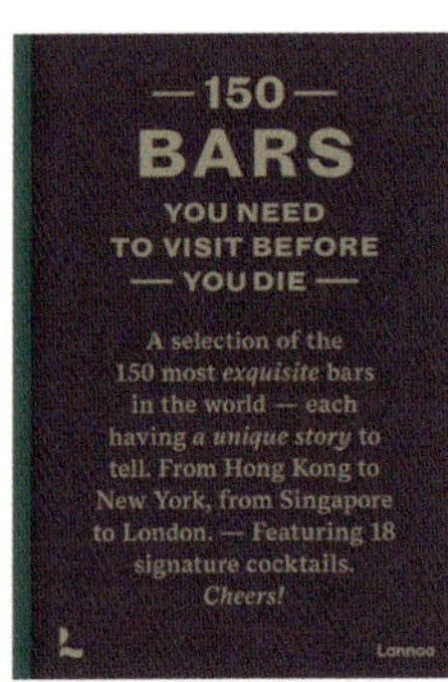

150 Bars
You Need to Visit
Before You Die
ISBN 9789401486194

150 Wine bars
You Need to Visit
Before You Die
ISBN 9789401486224

150 Bookstores
You Need to Visit
Before You Die
ISBN 9789401489355

Colophon

Texts
Stefanie Waldek

Copy-editing
Amy Haagsma

Book Design
ASB (Atelier Sven Beirnaert)

Sign up for our newsletter with news about new and forthcoming publications on art, interior design, food & travel, photography and fashion as well as exclusive offers and events. If you have any questions or comments about the material in this book, please do not hesitate to contact our editorial team: art@lannoo.com

D/2021/45/321 - NUR 450/500
ISBN 9789401479295
Third print run

www.lannoo.com